Gender, Tourism Entrepreneurship and Social Policy

This timely volume is a novel and important contribution to scholarly literature on gender and tourism entrepreneurship, utilising feminist and post-colonial frameworks to interrogate the role of social policies in facilitating inclusive tourism entrepreneurship.

Drawing on contributions and case studies from across the Global South and Global North, this multi-disciplinary collection identifies how regional variations in governance and policy influence the experiences and potentialities of tourism entrepreneurship as a promised avenue for inclusive growth for marginalised identities. Problematizing universalised constructions of entrepreneurs as necessarily masculine, western, and driven only by economic imperatives that seek to fix and dislocate entrepreneurial support, this volume takes focus with place-based approaches to explore the intersections between identity, tourism entrepreneurship and social policy. It is this geographically informed perspective that seeks to account for the complexity of entrepreneurial experience, and the role of social policy within this, that constitutes an original contribution to the field. The focus on gender and social policy reflects the increasing importance of tourism entrepreneurship within the context of the UNWTOs Sustainable Development Goals.

This book will be a pivotal resource for students, researchers, academics and policy makers in tourism, gender studies, development studies, sustainability and business.

Anna de Jong is a Senior Lecturer in Tourism in the College of Social Sciences in the School of Social and Environmental Sustainability at the University of Glasgow, United Kingdom. Anna is the Director of Research, within the School of Social & Environmental Sustainability. Anna is an interdisciplinary social scientist, with a background in human geography and tourism management. Her research takes focus with the relationships between tourism and place, guided by wider concerns of inequality and accessibility.

Michael Zisuh Ngoasong is a Professor of Entrepreneurship in the School for Business and Society at the University of York, UK; a Visiting Professor at The Open University, UK, where he is affiliated with the Centre for Social and Sustainable Enterprise; and a Senior Research Associate in the School of Hospitality and

Tourism at the University of Johannesburg, South Africa. Michael is the author of *Women's Entrepreneurial Journeys in Sub-Saharan Africa* (2023).

Albert Nsom Kimbu is a Reader and Head of the Tourism and Transport Department in the School of Hospitality and Tourism Management at the University of Surrey Guildford, UK. He is also a Senior Research Associate in the School of Tourism and Hospitality at the University of Johannesburg, Johannesburg, South Africa. Albert consults and has published widely on topics related to gender, entrepreneurship and tourism for equitable and inclusive development. He is the co-editor of *Routledge Handbook on Gender in Tourism: Views on Teaching, Research and Praxis* (Routledge, 2024).

Current Developments in the Geographies of Leisure and Tourism
Series Editors: Jan Mosedale and Caroline Scarles

Tourism and leisure exist within an inherently dynamic, fluid and complex world and are therefore inherently interdisciplinary. Recognising the role of tourism and leisure in advancing debates within the social sciences, this book series is open to contributions from cognate social science disciplines that inform geographical thought about tourism and leisure. Produced in association with the Geographies of Leisure and Tourism Research Group of the Royal Geographical Society (with the Institute of British Geographers), this series highlights and promotes cutting-edge developments and research in this field. Contributions are of a high international standard and provide theoretically informed empirical content to facilitate the development of new research agendas in the field of tourism and leisure research. In general, the series seeks to promote academic contributions that advance contemporary debates that challenge and stimulate further discussion and research both within the fields of tourism and leisure and the wider realms of the social sciences.

Moral Encounters in Tourism
Edited by Mary Mostafanezhad and Kevin Hannam

Travel and Imagination
Edited by Garth Lean, Russell Staiff and Emma Waterton

Neoliberalism and the Political Economy of Tourism
Edited by Jan Mosedale

Creating Heritage for Tourism
Edited by Catherine Palmer and Jacqueline Tivers

Gender, Tourism Entrepreneurship and Social Policy
Edited by Anna de Jong, Michael Zisuh Ngoasong, and Albert Nsom Kimbu

For more information about this series, please visit www.routledge.com/ Current-Developments-in-the-Geographies-of-Leisure-and-Tourism/book-series/ASHSER1372

Gender, Tourism Entrepreneurship and Social Policy

Edited by
Anna de Jong, Michael Zisuh Ngoasong, and Albert Nsom Kimbu

Routledge
Taylor & Francis Group
LONDON AND NEW YORK

First published 2025
by Routledge
4 Park Square, Milton Park, Abingdon, Oxon OX14 4RN

and by Routledge
605 Third Avenue, New York, NY 10158

Routledge is an imprint of the Taylor & Francis Group, an informa business

British Library Cataloguing-in-Publication Data
A catalogue record for this book is available from the British Library

ISBN: 978-1-032-32613-9 (hbk)
ISBN: 978-1-032-32614-6 (pbk)
ISBN: 978-1-003-31585-8 (ebk)

DOI: 10.4324/9781003315858

Typeset in Times New Roman
by codeMantra

Contents

Figures

Tables

Contributors

Aldo Alvarez-Risco is a fulltime researcher at Facultad de Ciencias Empresariales y Economía, Carrera de Negocios Internacionales, Universidad de Lima, Peru. He is also a pharmacist, has obtained a Doctor in Pharmacy and also has researched and published widely in health, green entrepreneurship and circular economy and sustainability.

Sara Calvo Martinez a Senior Lecturer in Entrepreneurship in the Department of Public Leadership and Social Enterprise at The Open University, UK. She is a co-founder of the social enterprise, Minca Ventures, an organisation that works to create impact and sustainability projects where they carry out consulting, coaching and training work. She is also the Founder of the Buen Vivir Institute, a social enterprise company that offers online education courses and coaching services for people and organisations.

Almudena Otegui Carles is an Associate Professor in the Facultade de Ciencias Empresariais e Turismo at Universidade de Vigo, Galicia, Spain. She is pursuing a doctorate in Tourism at the Universidade de Vigo in Spain. Almudena is the Guest Editor and Editorial Board Member for *Tourism Management Perspectives* (TMP) and of the Editorial Review Board of *International Journal of E-Business Research* (IJEBR).

Anna de Jong is a Senior Lecturer in Tourism in the College of Social Sciences in the School of Social and Environmental Sustainability at the University of Glasgow, United Kingdom. Anna is the Director of Research, within the School of Social & Environmental Sustainability. Anna is an interdisciplinary social scientist, with a background in human geography and tourism management. Her research takes focus with the relationships between tourism and place, guided by wider concerns of inequality and accessibility.

Shyla Del-Aguila-Arcentales, Escuela Nacional de Marina Mercante "Almirante Miguel Grau", Callao, Peru. Shyla has researched and publishes widely in green entrepreneurship and circular economy and sustainability.

José A. Fraiz-Brea is a Professor of Marketing in the Facultade de Ciencias Empresariais e Turismo at Universidade de Vigo, Galicia, Spain.

Albert Nsom Kimbu is a Reader and Head of the Tourism and Transport Department in the School of Hospitality and Tourism Management at the University of Surrey Guildford, UK. He is also a Senior Research Associate at the School of Tourism and Hospitality at the University of Johannesburg, Johannesburg, South Africa. Albert consults and has published widely on topics related to gender, entrepreneurship and tourism for equitable and inclusive development. He is the co-editor of *Routledge Handbook on Gender in Tourism: Views on Teaching, Research and Praxis* (Routledge, 2024).

Gizem Kutlu is a Lecturer in Business and Management at Coventry University London, UK, where she teaches undergraduate and postgraduate courses on Entrepreneurship and Innovation. Gizem recently completed a PhD at The Open University Business School, UK. Her research and teaching interests have a primary focus on the intersection of environmental entrepreneurship, sustainability and gender.

Sandy MacDonald is a Senior Lecturer in the Faculty of Business and Law at the University of Northampton, UK. Her research areas include gender and tourism and qualitative methods, and she has had a long involvement in higher education in Central Asia through work with the British Council and the Erasmus + programme.

Rehema Mugendi-Kiarie is a Lecturer in Marketing, Management, Digital Marketing and Sales at Riara University, Nairobi, Kenya, and a PhD student in Business Administration (Marketing option) at the University of Nairobi, Kenya. She has published widely around marking, including ethics in mobile banking, technology and social media use in customer service management and the development of marketing and selling of ethical products.

Michael Zisuh Ngoasong is a Professor of Entrepreneurship in the School for Business and Society at the University of York, UK; a Visiting Professor at The Open University, UK, where he is affiliated with the Centre for Social and Sustainable Enterprise; and a Senior Research Associate in the School of Hospitality and Tourism at the University of Johannesburg, South Africa. Michael is the author of *Women's Entrepreneurial Journeys in Sub-Saharan Africa* (2023).

Andrés Morales Pachón is a freelance research collaborator at The Open University, UK, and Goldsmith University, UK. Andres consults and has published widely on topics related to social enterprise and social innovation in European, African, Asian and Latin American countries. He is also a co-founder and COO of Minca Ventures, a social impact enterprise that offers training and educational programmes as well as consultancy and coaching.

Akbota Saduakassova is a master's student in the gender studies program at Corvinus University, Budapest. In 2020–2021, Akbota was a research fellow of the Gender Economics Centre at NARXOZ University, Kazakhstan, where she examined women's issues in the informal economy.

Aliya Tankibayeva is a Senior Lecturer at al-Farabi Kazakh National University, Department of Recreational Geography and Tourism, Kazakhstan. Her research interests are sustainable tourism to nature-based sites, gender and tourism, tourism and rural development. She has co-authored journal articles and analytical papers on sustainable natural resources management in regional development, including those from a comparative perspective.

Noelia Araújo Vila is a Lecturer in the Facultade de Ciencias Empresariais e Turismo at Universidade de Vigo, Galicia, Spain. Noelia is a member of the research group of the University of Vigo Emitur (Marketing and Tourism) and has published widely around tourism, marketing and business management.

Jaime A. Yáñez, Universidad Norbert Wiener, Vicerrectorado de Investigación and Teoma Global Gerencia Corporativa de Asuntos Científicos y Regulatorios, Lima, Peru.

1 Introduction to Gender, Tourism Entrepreneurship and Social Policy

Anna de Jong, Michael Zisuh Ngoasong and Albert Nsom Kimbu

Introduction

This collection is a novel contribution to the current conversations related to gender and tourism entrepreneurship and interrogating the role of social policies in facilitating inclusive tourism entrepreneurship. Tourism entrepreneurship has been identified as fundamental to meeting the UN's Sustainable Development Goals of gender equality (SDG 5), decent work and economic growth (SDG 8) and sustainable consumption and production (SDG 12); but as Figueroa-Domecq, de Jong and Williams (2020, 1) note, "neither entrepreneurship nor sustainability are gender neutral in the tourism industry", whilst engagement with policy is notably absent within research on gender and tourism entrepreneurship (Figueroa-Domecq, de Jong & Williams, 2020a). It is recognised that the tourism sector provides women and other disadvantaged groups (e.g., indigenous groups, LGBTQ communities) with employment and entrepreneurial opportunities, both vital for their economic and social independence. For women in these groups (and other disadvantaged entrepreneurs), their tourism businesses also act as platforms for the promotion of local development and social transformation, enabling them to create 'new' self-identities (Kimbu & Ngoasong, 2016).

Drawing on in-depth, situated case studies from across the Global South and Global North, this multi-disciplinary collection identifies how regional variations in governance and policy influence the experiences and potentialities of tourism entrepreneurship as a promised avenue for inclusive growth for marginalised identities. Problematising universalised constructions of entrepreneurs as necessarily masculine, western, and driven only by economic imperatives that seek to fix and dislocate entrepreneurial support, this collection takes focus on place-based approaches to explore the intersections between identity, tourism entrepreneurship, and social policy. It is this geographically informed perspective that seeks to account for the complexity of entrepreneurial experience, and the role of social policy within this, that constitutes an original contribution to the field. All contributions to the collection will be original, and authors will not be allowed to replicate work that was previously published. A double peer-review process aims to ensure the academic credibility of the chapters included in the collection. The editors and contributors present the broad demographics of scholars in tourism.

DOI: 10.4324/9781003315858-1

COVID-19 has exacerbated the multiple challenges faced by all entrepreneur owner/managers of tourism firms, underscoring the imperative role social policy undertakes in supporting tourism entrepreneurs. The COVID-19 situation is even worse for developing and emerging destinations of the Global South, where the majority of tourism micro, small, and medium enterprises (MSMEs) are owned by women and members of disadvantaged groups who, unlike tourism and hospitality businesses of the Global North, have had little-to-no support from their national governments leading to the closure of many tourism and hospitality businesses, placing many millions of jobs at risk and threatening to roll back the progress made in sustainable development (Rogerson & Baum, 2020; UNWTO, 2020). Consequently, the UNWTO (2020) estimates that destinations (especially in the Global South), which are for the most part dependent on international visitors, would be the worst hit and the slowest to recover post-COVID-19. The crisis has laid bare the precarity of entrepreneurial operations of the different actors especially those operating MSMEs who were not prepared for such crisis and lack the resilience to respond and overcome it, highlighting the crucial role social policy plays in ensuring the sustainability of entrepreneurs.

Conceptualising and Linking Gender, Entrepreneurship, and Social Policy in Tourism

To produce knowledge that can influence policy and social change in tourism, we see a gap in combining disciplinary and interdisciplinary research at the intersection of gender, entrepreneurship, and social policy as critical. Although entrepreneurship leads to equality (Vracheva & Stoyneva, 2020) and increasing gender awareness can improve social policy (Alarcón & Cole, 2019), these three fields have not been studied together. Our aim in this book is to define and establish a distinctive new field integrating gender, entrepreneurship, and social policy (Figure 1.1). To this book, we identify tourism as integral in these interlinkages because it is often positioned as an avenue for inclusive growth for marginalised identities in a broad range of international contexts. In taking a place-based approach that centres tourism, this collection is thus able to explore the complexity of entrepreneurial experience and the role of social policy within this. This includes not only building, nationally and internationally, on the critical linkages between these distinctive areas of study but also comparing different specialist industries, geographic destinations, and localities. In this section, we will set out what is known about conceptualisation and interlinkages between gender, entrepreneurship, and social policy and how this book seeks to fill existing gaps in the literature.

Gender and Social Policy

Sinclair's (1996) seminal collection on gender, work, and tourism discusses the central role played by women and the nature of their work in the tourism industry. The book covers a wide geographical range and contributes to the study of gender and work. It focuses only on work and gender in tourism and mostly focuses on the

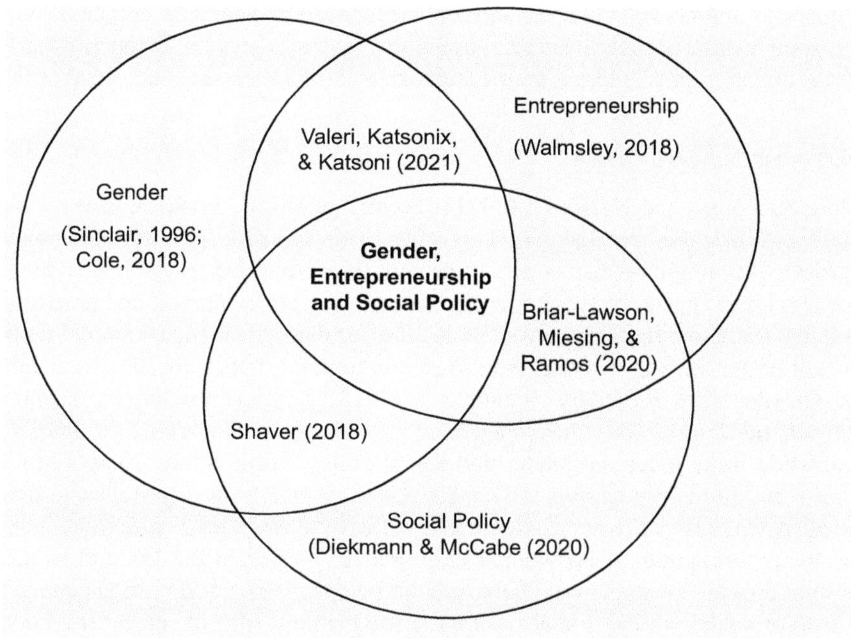

Figure 1.1 Gender, entrepreneurship, and social policy linkages

aspects of women. Other aspects of gender studies are not discussed in this book, whilst the role of policy and entrepreneurship is omitted. A more recent collection focuses on gender equality in tourism and argues the case for researchers to be more critical in their treatment of the extent to which tourism brings women's empowerment (Cole, 2018). In particular, the articles in the edited collection by Cole (2018) emphasise an approach that integrates multiple stakeholder perspectives, from INGOs, national governments, and managers, as well as workers in a variety of fields that produce and benefit from tourism. The contribution of this work is its focus on women's empowerment. Though some of the chapters touch on entrepreneurship, they do not engage explicitly with it, and the book does not explore the experiences of gender beyond that of 'women'.

In terms of social policy, Diekmann and McCabe's (2020) *Handbook of Social Tourism* considers the importance of emphasising the impacts and challenges of tourism on people's lives from a social perspective whilst considering the potential of tourism as a force for good. Shaver's (2018) *Handbook on Gender and Social Policy* meanwhile covers gender inequalities of income, employment and care, and family policy and examines gender and social policy at both the regional and national levels. The book however does not engage with tourism or entrepreneurship. Wilson and Chambers (forthcoming) builds on the extant literature to continue advancing the research agenda for gender and tourism. The editors argue that research has not fully exposed and dismantled the historic and structural inequalities, human right abuses, and injustices faced by people of colour. Though embracing a different, empowered, more loving, and more equitable tourism world, there is no intention to address themes related to entrepreneurship and social policy.

Gender and Entrepreneurship

Valeri, Katsonix, and Katsoni's (2021) recently published book on *Gender and Tourism: Challenges and Entrepreneurial Opportunities* adopts a multi-disciplinary approach, building on a historically informed, future-focused research agenda that accounts for the needs and concerns of contemporary policy makers and practitioners in the tourism field. The book contains limited theoretical engagement with the concept of gender and is broad in its focus on tourism, with only limited insights into social policy. Relatedly, Walmsley's (2018) *Entrepreneurship in Tourism: A Practical Guide to the Development of a Business for Tourism Entrepreneurs* meanwhile lacks focus on gender and social policy. Furthermore, Elam's (2008) *Gender and Entrepreneurship: A Multilevel Theory and Analysis* provides a sociological perspective on why women are less likely to undertake entrepreneurship. The focus throughout is on women entrepreneurs based in the US and as such does not provide insights into diverse gender perspectives, nor does it engage with tourism or social policy. It is also important to acknowledge the established body of research around women's entrepreneurship. A collection by Yousafzai, Fayolle, Lindgreen, Henry, Saeed, and Sheikh (2018) focuses on the performance of women-owned businesses and how performance is defined through the institutional power structures and policies. The book challenges scholars to rethink how to overcome constrained performance and encourage women's entrepreneurial activities through a variety of methodological approaches and geographical contexts.

Entrepreneurship and Social Policy

Briar-Lawson, Miesing, and Ramos's (2020) book *Social Entrepreneurship and Enterprises in Economic and Social Development* offers a theoretical perspective around social entrepreneurship that focuses on showing how enterprises can integrate social and economic development missions in their strategy and operational activities. The editors argue that the pursuit of a dual mission that strives to achieve both financial sustainability and social good is especially path-breaking approaches to delivering the UN SDGs. The chapters of the book reveal social policies that promote support for social entrepreneurs since environmental, economic, and social sustainability are core goals, whilst challenging the extent to which fostering social enterprises is the panacea to addressing human needs when government investments are required in social welfare, social protections, and ecosystem supports. Whilst broad in focus, none of the chapters explicitly discuss tourism.

A related collection by Diaz-Garcia, Brush, Gatewood, and Welter (2016) puts women's entrepreneurship in global and local contexts, with context viewed from several dimensions (historical or temporal, geographic or spatial, institutional, social, organisational, and cultural). Several chapters of this book cover strategies and policies influencing entrepreneurial start-up decisions of women, gendered contextualisation of SME cooperation in urban versus rural setting, and community relations strengthening through social entrepreneurship. We see opportunities to extend the conversation around organisational policies and strategies to include an explicit focus on social policy influences that encourage and challenge organisational strategies in a range of context.

Researching the Management of Tourism in Developing and Developed Economies

Though entrepreneurship underpinned by good social and economic policies leads to equality, inclusion, and increasing gender awareness, these three fields have not been studied together within the context of tourism. Extant studies have addressed gender and women entrepreneurship in tourism with findings explicating a higher degree of environmental and social responsibility among women (Figueroa-Domecq, Kimbu, et al., 2020), the lack of skills (Kimbu, Ngoasong, Adeola & Afenyo-Agbe, 2019), access to financial and nonfinancial resources (Kimbu & Ngoasong, 2016; Ngoasong & Kimbu, 2019), infrastructural deficiencies (UNWTO, 2020), discrimination (Ahl & Marlow, 2012), political, social and cultural inhibitions (Zhang, Kimbu, Lin & Ngoasong, 2020), all of which have been found to limit women and other marginalised groups from reaching their full business potential and subsequent contribution to societal change. Building on these studies, the current book collectively investigates the nexus between gender, entrepreneurship, and social policy in a way that recognises the complex nature of this relationship (Ahl & Marlow, 2021) and the imperative role social policy will play in the COVID-19 response.

Linking gender, entrepreneurship, and social policy introduces the problems of access to products/services and reach (e.g., poor or marginalised communities). Demographic characteristics such as gender identity, religion, and race can restrict access to products and services for certain groups whilst market inefficiency and injustices can reduce welfare, all of which create challenges for organisations working on sustainability (George, Merrill & Schillebeeck, 2020). Consequently, critical (re)examination of social policies, gender, and entrepreneurship-related practices within tourism from a sustainability perspective is warranted. We are particularly interested in research that engages with critical feminist and post-colonial perspectives that interrogate the roles of (social) policies and practices in promoting inclusion and equality through tourism. Incorporating critical social theory into research on gender, entrepreneurship, and social policy renders insights into how gender is a social ascription performed by individuals in various ways, depending on the context (Figueroa-Domecq et al., 2020). Examining how gender is constructed through entrepreneurship and social policy allows us to question the settlement of 'truths' and begin to (re)consider how gender might be differently conceptualised. The incorporation of critical social theory therefore goes hand in hand with the diverse perspectives each chapter offers – with each chapter exploring one or more economic, social, cultural, political, organisational, or environmental dimensions of the subject from national, cross-national, and global contexts.

Research on gender, entrepreneurship, and social policy will also benefit from quantitative and qualitative methodologies, as well as mixed methods (Van der Duim, Lamers & van Wijk, 2015). Research on tourism development is characterised by the widespread adoption of qualitative methods, many of which are participatory. A wide range of methods, including the use of visual, photographic, and participatory methods as well as the more standard in-depth interviews and focus group discussions, are commonly used in the literature (Richards & Munsters, 2010; Rakic & Chambers, 2013). This book follows in this tradition by bringing together diverse methodological approaches. The difficulties of conducting research that draws on illustrative cases and representations from across the world are well documented and have contributed to the dearth of up-to-date findings. Bakas (2018), by way of example, brought attention to the challenges in undertaking in-depth ethnographic research with women entrepreneurs because femininity is connected to 'being constantly busy'.

Whilst Henry, Foss, and Ahl (2015) implies that the very cultural, legislative, and economic frameworks that present gendered challenges for some entrepreneurs also present barriers for those seeking to research underrepresented groups because such entrepreneurs from these groups are more likely to be invisible and inaccessible. Such challenges perhaps influence Figueroa-Domecq et al.'s (2020) finding that the lowest levels of research around gender and tourism entrepreneurship come from Central and South America, followed by Africa and Asia. To ensure a representative sample of countries and geographic coverage, the chapters have been selected to reflect geographical regions currently underrepresented. The contributors to this volume are interdisciplinary, consisting of scholars from developed and developing countries across five continents.

Structure and Contributions of Book

This book contains ten chapters covering the over-arching framework that positioned the intersection of gender, tourism entrepreneurship, and social policy (Figure 1.1) in Chapter 1. Chapters 2–9 address different aspects of this intersection and associated implications in the context of contrasting tourism geographies. Together the thematic areas can be grouped into three broad areas.

First, the persistent gendered challenges that necessitate the gendering of social policy in ways that encourage more entrepreneurship. This is illustrated in contrasting geographic contexts, to gain a sense of the various ways through either which social policy has been found to support tourism entrepreneurship (Chapter 2) or where social policy can address historical gendered challenges facing women entrepreneurs (Chapter 3). Chapter 2 presents the experience on rural tourism entrepreneurship and social policy in the Latin American context of Peru. Chapter 3 uncovers the gendered challenges faced by a women entrepreneur when creating new entrepreneurial identity and establishing legitimacy in Turkey. A common theme across the two chapters, however, is that gender is largely omitted from social policy, despite gender playing an influential role in women's entrepreneurial journeys and how governments ought to be supporting tourism entrepreneurship and intrapreneurship.

Second, the gendered aspects of varieties of tourism entrepreneurship and the role of social policy in facilitating entrepreneurial leadership and networking. By varieties of tourism entrepreneurship, we are referring to entrepreneurial initiatives in women-owned businesses and how organisations can be entrepreneurial in their support for gender equality. Through a case study of "Las Kellys" in Spain, Chapter 4 uncovers how hotel housekeepers can be entrepreneurial in confronting precariousness through formal associations known as "Las Kellys", creating collective resistance in the struggle for their rights. Chapter 5 uncovers women-entrepreneurs sense-making of policy support in context of rural tourism in Kazakhstan. Chapter 6 analyses the external tourism environment within which social policies for supporting women in the tourism industry can be best understood in Kenya. A central theme across the chapters that address this theme is that social policy has tended to ignore entrepreneurial leadership and networks. Therefore, there is a need to clarify decisions about social policy and mobilising key actors to support women's tourism entrepreneurs. The central actors within entrepreneurial networks are explored, including government agencies, unions, business associations, and grassroots organisations.

Third, the policy practice implications of financing for women's tourism entrepreneurship. As identified by Figueroa-Domecq et al. (2020a), there has been limited research assessing the gendered dimensions of entrepreneurial training and financing, an oversight considering the breadth of social policies that present both training and financing as ways through which to support tourism entrepreneurs across genders, enterprise types, and geographical locations. In consideration, this section takes focus on the assessment of financing initiatives. Chapter 7 critically discusses the gendered dynamics of entrepreneurial finance in women's tourism entrepreneurship, using case studies developed in Nigeria and Ghana. Chapter 8 presents an alternative route to financing tourism entrepreneurship through a study "Buen Vivir" for indigenous tourism entrepreneurship in Columbia. Both these chapters reveal how the combination of formally regulated and informal financing is needed to promote women's tourism entrepreneurship and the importance of the collective action found in local communities to enable women's role in tourism entrepreneurship. Chapters 9 argues that the barriers faced by female entrepreneurs before and during COVID-19, a major crisis, require novel framework for enabling resilience for sustainability in tourism entrepreneurship.

Finally, Chapter 10 provides our conclusions and reflections on the future of gender, tourism entrepreneurship, and social policy. This book reveals how gendered bias involved in tourism entrepreneurship favours masculine entrepreneurial performances as well as issues of costs and benefits faced by different stakeholders. Recognising existing biases and costs/benefits and the solutions offered by the chapters in the book, Chapter 10 critically examines the effectiveness of social policy, as complementary to other policy areas, such as financing, trade, and development, in enabling women in becoming 'successful' entrepreneurs in tourism development.

References

Ahl, H., & Marlow, S. (2012). Exploring the dynamics of gender, feminism and entrepreneurship: advancing debate to escape a dead end? *Organization, 19*(5), 543–562.

Ahl, H., & Marlow, S. (2021). Exploring the false promise of entrepreneurship through a postfeminist critique of the enterprise policy discourse in Sweden and the UK. *Human Relations, 74*(1), 41–68.

Alarcón, D., & Cole, S. (2019). No sustainability for tourism without gender equality. *Journal of Sustainable Tourism, 27*(7), 903–919.

Bakas, F. (2018). 'Crafting an entrance': gender's role in gaining and maintaining access in tourism ethnography and knowledge creation. In Andrews, H., Takamitsu, J. & Dixon, L. (Eds.), *Tourism Ethnographies: Ethics, Methods, Application and Reflexivity* (pp. 67–82). London: Routledge.

Briar-Lawson, K., Miesing, P., & Ramos, B. M. (2020). *Social Entrepreneurship and Enterprises in Economic and Social Development.* Oxford: University Press.

Cole, S. (Ed.). (2018). *Gender Equality and Tourism: Beyond Empowerment.* Wallingford: CABI.

Diaz-Garcia, C., Brush, C., Gatewood, E., & Welter, F. (Eds.). (2016). *Women's Entrepreneurship in Global and Local Contexts.* Cheltenham: Edward Elgar.

Diekmann, A., & McCabe, S. (2020). *Handbook of Social Tourism.* Research Handbooks in Tourism series. Cheltenham: Edward Edgar.

Elam, A. B. (2008). *Gender and Entrepreneurship: A Multilevel Theory and Analysis.* Cheltenham/Northampton, MA: Edward Elgar.

Figueroa-Domecq, C., de Jong, A., & Williams, A. M. (2020a). Gender, tourism and entrepreneurship: a critical review. *Annals of Tourism Research, 84*, 102980. https://doi.org/10.1016/j.annals.2020.102980.

Figueroa-Domecq, C., Kimbu, A., de Jong A., & Williams, A. M. (2020b). Sustainability through the tourism entrepreneurship journey: a gender perspective. *Journal of Sustainable Tourism.* https://doi.org/10.1080/09669582.2020.1831001

George, G., Merrill, R. K., & Schillebeeck, S. J. D. (2020). Digital sustainability and entrepreneurship: how digital innovations are helping tackle climate change and sustainable development. *Entrepreneurship Theory and Practice,* 45(1). https://doi.org/10.1177/1042258719899425

Henry, C., Foss, L., & Ahl, H. (2015). Gender and entrepreneurship research: a review of methodological approaches. *International Small Business Journal, 34*(3), 217–241.

Kimbu, A. N. & Ngoasong, M. Z. (2016). Women as vectors of social entrepreneurship. *Annals of Tourism Research, 60*, 63–79.

Kimbu, A. N., Ngoasong, M. Z., Adeola, O., & Afenyo-Agbe, E. (2019). Collaborative networks for sustainable human capital management in women's tourism entrepreneurship: the role of tourism policy. *Tourism Planning & Development, 16*(2), 161–178.

Ngoasong, M. Z., & Kimbu, A. N. (2019). Why hurry? The slow process of high growth in women-owned businesses in a resource-scarce context. *Journal of Small Business Management, 57*(1), 40–58.

Rakic, T., & Chambers, D. (Eds.). (2013). *An Introduction to Visual Research Methods in Tourism.* London/New York: Routledge.

Richards, G., & Munsters, W. (Eds.). (2010). *Cultural Tourism Research Methods.* Wallingford: CABI.

Rogerson, C. M., & Baum, T. (2020). COVID-19 and African tourism research agendas. *Development Southern Africa, 37*(5), 727–741.

Shaver, S. (2018). *Handbook on Gender and Social Policy*. Cheltenham: Edward Elgar.
Sinclair, M. T. (Ed.). (1996). *Gender, Work and Tourism*. London: Routledge.
UNWTO. (2020). *AlUla Framework for Inclusive Community Development through Tourism*. Madrid: United Nations World Tourism Organisation. https://doi.org/10.18111/9789284422159.
Valeri, M., Katsonix, V., & Katsoni, V. (2021). *Gender and Tourism: Challenges and Entrepreneurial Opportunities*. Bingley: Emerald Insight.
Van der Duim, R., Lamers, M., & van Wijk, J. (Eds.). (2015). *Institutional Arrangements for Conservation, Development and Tourism in Eastern and Southern Africa: A Dynamic Perspective*. Dordrecht: Springer.
Vracheva, V., & Stoyneva, R. (2020). Does gender equality bridge or buffer the entrepreneurship gender gap? A cross-country investigation. *International Journal of Entpreneurial Behaviour & Research, 26*(8), 1827–1844.
Walmsley, A. (2018). *Entrepreneurship in Tourism*. London: Routledge.
Wilson, E., & Chambers, C. (2023). *A Research Agenda for Gender and Tourism*. Cheltenham: Edward Elgar.
Yousafzai, S., Fayolle, A., Lindgreen, A., Henry, C, Saeed, S., & Sheikh, S. (Eds.). (2018). *Women Entrepreneurship and the Myth of Underperformance: A New Look at Women's Entrepreneurship Research*. Cheltenham: Edward Elgar.
Zhang, C. X., Kimbu, N. A., Lin, P., & Ngoasong, M. Z. (2020). Guanxi influences on women intrapreneurship. *Tourism Management, 81*, 104137. https://doi.org/10.1016/j.tourman.2020.104137

2 Tourism, Gender Entrepreneurship and Social Policy

A Case Study of Peru

Shyla Del-Aguila-Arcentales, Aldo Alvarez-Risco and Jaime A. Yáñez

Introduction

Peru, as a pluricultural (Ombudsman of Peru, 2023) and megadiverse (World Bank, 2013) country, possesses a great material and immaterial wealth that makes it one of the main tourist destinations in the world (Cambra-Fierro et al., 2022; Carvache-Franco et al., 2022, 2021; Gabriel-Campos et al., 2021; Pérez Gálvez et al., 2017; Ruiz-Ortega et al., 2021). These attributes attract thousands of tourists year after year and have provided sources for survival to the different communities of the Peruvian territory. Peru is a world-renowned tourist destination that captivates travellers from all over the world with its rich history, stunning landscapes, and vibrant culture. Located in the western part of South America, this diverse country offers a unique experience that combines the majesty of the Andes, the mysterious Amazon rainforest, and the fascinating Pacific coast.

Table 2.1 shows the awards of Peru in tourism by World Travel Awards (www.worldtravelawards.com), being relevant award to Peru as culinary and tourism destination.

With a history dating back thousands of years, Peru is home to ancient civilizations such as the Incas, who left an extraordinary cultural legacy in the form of majestic ruins, such as the citadel of Machu Picchu (Cusco Peru, 2023), which stands high in the Andes and is considered one of the seven wonders of the modern world. In addition to Machu Picchu, Peru is home to a host of archaeological sites, such as the enigmatic Nazca lines (National Grographic, 2023) and the ruins of the city of Cusco, which served as the capital of the Inca empire.

Method

To conduct this study, we have used a data collection strategy involving various reliable and relevant sources of information in the field of female entrepreneurship and tourism. We have consulted the scientific literature, which provides a solid foundation based on previous research and theories developed by experts in this area. The scientific literature offers a comprehensive perspective on the advancements, challenges, and opportunities in female entrepreneurship in the tourism

DOI: 10.4324/9781003315858-2

Table 2.1 Peru in World Travel Awards (2012–2023)

Winner	*Year*
World's leading culinary destination	
Peru	2012–2023
World's leading cultural destination	
Peru	2018–2019, 2021–2023
World's leading tourist attraction 2017	
Peru	2017–2019, 2021–2022
South America's leading business hotel	
JW Marriott Hotel Lima, Peru	2021, 2023
Hilton Lima Miraflores, Peru	2022
Swissôtel Lima, Peru	2013–2015, 2019
South America's leading green hotel	
Tambopata Research Center, Peru	2020–2023
South America's responsible tourism award	
Llama Path, Peru	2022–2023
Posada Amazonas Lodge, Peru	2020–2021

sector, aiding in a better understanding of the influencing context and dynamics. In addition to the scientific literature, we have utilized official websites that present national programmes and specific laws related to women's entrepreneurship in tourism. These websites include government and international organization portals, which often publish up-to-date and relevant information on political and legal initiatives designed to promote and support female entrepreneurship. Consulting these official resources is crucial as they offer precise data on the regulatory framework, current public policies, and available support and financing programmes for women entrepreneurs in the tourism sector.

The combination of these sources allows us to obtain a comprehensive and updated vision of the state of female entrepreneurship in tourism, considering both academic knowledge and policies and programmes applied in practice. This methodology ensures that the study is based on solid and diverse evidence, covering different aspects of the topic investigated. The primary data repository used to obtain information for this study is Scopus. Scopus is one of the largest and most comprehensive databases of peer-reviewed scientific literature, known for its comprehensive coverage of various academic and scientific disciplines. The choice of Scopus is due to its breadth and depth in the collection of scientific articles, conferences, and other types of academic documents, ensuring access to a vast amount of relevant and up-to-date research. Using Scopus as a primary source allows researchers to access high-quality studies and detailed and specific information on female entrepreneurship in tourism. Additionally, Scopus facilitates the identification of research trends, tracking citations, and evaluating the impact of publications, which is essential for developing rigorous and informed analysis.

Results

A fundamental characteristic is that the entrepreneurs, who can be individuals or legal entities, promote a business linked to tourism-related services or products, which may include tour guides for visitors, handicrafts that can be sold to visitors, and lodging and food for tourists, among others. Another crucial aspect is that these entrepreneurs develop a local plan that ensures the sustainability of tourism in that locality, thus avoiding the overexploitation of natural resources or the deterioration of a rural area of tourist interest. After the COVID-19 pandemic, Peru began to have an arrival of tourists in almost the same number as before the pandemic. In this way, Peru registered more than 2.5 million international tourists in 2023 (Government of Peru, 2024).

Social policies have aimed to support women entrepreneurs in the tourism sector. This is in recognition of the need to promote gender equality and aid in the economic recovery of an industry deeply affected by the pandemic. Various strategies have been put in place to empower women and ensure they have access to the opportunities necessary to thrive in tourism. One such strategy is the development of specific training programmes for women entrepreneurs in tourism. These programmes focus not only on technical and business management skills but also on the development of digital competencies, which are proving to be essential in the post-pandemic era. The digitalization of tourism is a growing trend, and training women in the use of digital tools can help them attract a broader and more diversified clientele. This shift towards digitalization offers a hopeful future for women in the tourism sector as it opens up new avenues for success and growth. Examples of these training programmes include courses in digital marketing, social media management, and e-commerce, which are crucial for success in an increasingly globalized and digital market.

In addition, social policies have promoted access to financing for women entrepreneurs in the tourism sector. Traditionally, women face greater difficulties in accessing credit and financing due to structural barriers and discrimination. Specific funds and lines of credit have been created to support women in tourism, offering favourable conditions and low interest rates. This financial support is vital so that entrepreneurs can reactivate or start their businesses, make necessary investments, and maintain their operations during periods of low demand. Another key strategy has been implementing policies that encourage women's participation in decision-making within the tourism sector. This includes incentives for more women to occupy leadership positions in tourism businesses and tourism-related government organizations. By increasing female representation at decision-making levels, we seek to ensure that women's perspectives and needs are considered in the planning and development of tourism policies.

Likewise, work has been done to create support and mentoring networks for women entrepreneurs. These networks provide a space to share experiences, challenges, and solutions, facilitating mutual learning and growth. Mentoring can offer crucial support, providing guidance and practical advice from experienced women in the sector. Finally, social policies have sought to address structural inequalities that affect women in tourism, such as the wage gap and job insecurity. Measures such as promoting fair employment contracts and protecting against workplace harassment and discrimination are essential to creating a safe and equitable work environment for women.

During the COVID-19 pandemic the tourism industry was impacted seriously (Collins-Kreiner & Ram, 2021; Sigala, 2020). In this way, a report, published by the United Nations World Tourism Organization (UNWTO) and the United Nations Conference on Trade and Development (UNCTAD), states that international tourism, and its highly linked sectors, suffered an estimated loss of $2.4 trillion in 2020 after a sharp drop in international tourist visits (UNCTAD, 2021) and also because tourist destinations recorded 1 billion fewer international arrivals in 2020 than in 2019 (World Bank Blog, 2021). Peru has also been affected by this situation so during the pandemic, businesses had to adapt by creating new business models (RPP, 2021). The biggest challenges faced by the tourism industry were to regain the trust of tourists, promote tourist destinations, design new tourism products, and create and implement tools for collecting and processing relevant data for managing destinations and making decisions for public institutions.

To address this situation, the Peruvian government established the National Strategy for the Reactivation of the Tourism Sector 2022–2025 (Government of Peru, 2021), which detailed some initiatives to boost tourism activity in Peru that are aligned with key issues: Economic resources by the creation of information systems looking the optimization of international relations for technical and financial cooperation. Then, the improvement of conditions of tourism in destinations by creating adequate conditions for tourism service providers. The connectivity for passenger traffic was an important issue established to promote better conditions of the environment for tourists.

Social policies play a fundamental role in the development of tourism entrepreneurship in rural areas of Peru, especially when they are led by women. In a country as rich in cultural and natural diversity as Peru, tourism is presented as an invaluable opportunity for economic growth and social inclusion, especially in remote and marginalized regions (MIDAGRI, 2022). However, for this potential to be fully realized, the support of social policies that promote gender equality, access to education and training, and economic support for female entrepreneurs is required. Peruvian women play a crucial role in promoting tourism in rural areas since they are not only guardians of ancestral traditions and custodians of biodiversity, but they are also agents of change and development in their communities (Andes Resilient to Climate Change, 2022; Rainforest Alliance, 2022). These traditions cover a wide range of activities, including textile making, agriculture, gastronomy, and local festivities.

Weaving techniques, for example, are not only a testament to skill and creativity but also represent a link to the history and cultural identity of their communities. Women weave stories into each piece of fabric, using patterns and colours that have specific meanings and that reflect their worldview and relationship with nature. Women play a crucial role in conserving local biodiversity. They are responsible for agriculture and collecting medicinal plants, and they apply sustainable practices that have been refined over centuries. These practices not only ensure the survival of their families but also help preserve local ecosystems. For instance, their use of traditional agricultural techniques like crop rotation and saving native seeds helps maintain soil health and the genetic diversity of cultivated plants. Women play a crucial role as custodians of traditions, including organizing community celebrations and rituals. These events are essential for fostering social unity and passing down cultural values. Women often take the lead in planning and taking part in these occasions, ensuring that traditions endure and evolve with the times while preserving their essence. Dance, music, and religious ceremonies are central to these festivities, and women's active involvement guarantees that these customs remain an integral part of community life.

Social policies that promote the active participation of women in the tourism industry not only empower these entrepreneurs but also generate a positive impact on their families and society (Action Against Hunger, 2022). Empowerment has seen an increasing importance and has become one of the buzzwords. However, women's empowerment remains a central objective of international development (Mosedale, 2014). Empowerment was intended to transform gender inequality but has come to mean providing income to women (or perhaps even increasing the workforce and providing companies with cheap employees to exploit). Women need support to ensure that the empowering potential of entrepreneurship is realized, that is, materialized. Valuable debates within tourism entrepreneurship studies highlight the empowering potential of entrepreneurship, specifically for women (Figueroa-Domecq et al., 2020). Women entrepreneurs often rely on informal networking and may find informal financing more accessible than formal sources, especially in rural areas. Social policies should prioritize supporting the development of ventures that help people thrive, taking into account the significant role of women in these ventures (Kimbu et al., 2024).

Access to financing, training programmes in business management and tourism marketing, and the strengthening of support networks and associations are key elements that can be promoted through specific social policies. Furthermore, it is essential that these policies promote environmental and cultural sustainability, ensuring that tourism benefits both local communities and the environment (Abou-Shouk et al., 2021; Elshaer et al., 2021).

The chapter focuses on describing Peru's advances in entrepreneurship developed by women in tourism areas. The following text seeks to emphasize the significant role that women have played in the advancement and growth of tourism. They have not only been a source of economic income but have also contributed to the preservation and enhancement of local culture and heritage. The analysis intends to showcase the various initiatives and projects led by women that have diversified tourism offerings in different regions of the country, promoting sustainable and responsible tourism. Furthermore, the chapter explores the challenges and obstacles faced by women entrepreneurs in the tourism sector as well as the strategies and policies implemented to support and encourage their active participation. It also considers aspects such as access to financing, training, support networks, and the use of innovative technologies to facilitate the management and promotion of their ventures. The aim is to provide a comprehensive and detailed overview of the accomplishments and contributions of women in the tourism industry in Peru, emphasizing the significance of their empowerment and the positive impact of their initiatives on local communities and the national economy. This effort seeks to inspire and motivate more women to participate in the tourism sector, recognizing their potential as agents of change and sustainable development.

In 2007, a strategy called Rural Community Tourism (MINCETUR, 2007) was implemented, which is aimed at seeking social and economic inclusion in rural areas, including tourist areas of Peru. It is necessary to specify what is understood by Rural Community Tourism. Rural Community Tourism is understood, within the context of this strategy, as tourist activity that is developed in rural areas that has the participation of local populations in an organized way, seeking the benefit of families and where the focus of the tourist service is rural culture.

A fundamental characteristic is that the entrepreneurs, who can be individuals or legal entities, promote a business linked to tourism-related services or products, which may include tour guides for visitors, handicrafts that can be sold to visitors, and lodging and food for tourists, among others. Another crucial aspect is that these entrepreneurs develop a local plan which ensures the sustainability of tourism in that locality, thus avoiding the overexploitation of natural resources or the deterioration of a rural area of tourist interest.

Tourism is a significant industry that relies on nature as its raw material. However, the use of natural areas for tourism can cause environmental degradation, leading to the failure of tourist destinations. The situation is especially concerning because popular tourist destinations are often fragile ecosystems with a wide variety of species (Priatmoko et al., 2023; Weyland et al., 2021).

The unsustainable use of these areas can lead to irreversible losses of renewable and non-renewable energies. Despite these issues, tourism can have positive impacts on a country's economy, society, politics, and culture. Rural tourism is a type of tourism that occurs in rural or countryside areas, often in small towns or villages. Its primary goal is to provide tourists with an authentic experience in proximity to nature, culture, and local traditions. Rural tourism has become increasingly popular in recent years as travellers seek more authentic experiences and contact with nature (Che et al., 2021; Kumar & Shekhar, 2020). Rural tourism is an excellent alternative to the over-tourism that occurs in urban and coastal locations. It boosts economic and social development and preserves cultural and natural heritage in these areas.

Rural tourism has become an increasingly popular choice for those interested in learning about the culture and everyday lives of community members and indigenous peoples in rural areas. It provides interesting experiences, promotes sustainability, and preserves cultural and natural heritage (Coroş et al., 2021; Rosalina et al., 2021). In Peru, rural tourism is mainly focused on the Andean and Amazon regions, where visitors can find communities with rich history, traditions, cuisines, and crafts. This type of tourism plays a critical role in the preservation and restoration of the basic elements that make it up, such as crafts, gastronomy, folklore, customs, and traditional activities. Cuzco, for example, is home to numerous archaeological sites, including the world-famous Machu Picchu (Fox, 2021). While most tourists visit Cuzco and its nearby historical places, rural tourism initiatives aim to create economic opportunities for locals while preserving their cultural heritage and natural environment. Homestays, in which visitors stay with local families and engage in their daily activities, and community-run tours that highlight traditional farming practices and handicrafts are among these initiatives. San Martin, a region in the country's northeastern region, is known for its captivating unique scenic and cultural heritage. Many of its towns have homestays where tourists can spend a night with villagers and participate in activities like weaving, fishing, and agriculture. The region also has several protected areas, such as the Cordillera Escalera Regional Conservation Area, which offers hiking, birdwatching, and other outdoor activities.

Problems of Rural Tourism in Peru

In Peru, rural tourism has become a significant economic activity over the past decade. It has helped families in rural areas to improve their living conditions by generating higher incomes (La Camara, 2020). Destinations like Cusco, Puno, and Arequipa have gained popularity among visitors who are interested in local culture, biodiversity, and gastronomy. Rural tourism in Peru has several advantages, such as higher employment rates, compared to other rural activities (OXFAM, 2024); however, it also faces various challenges as support of programmes and training in rural areas, including sociocultural, geographic, and security-related issues, among others. These challenges can impact not only the residents of these places but also the number of tourists who visit them.

The COVID-19 pandemic has been a significant challenge for the tourism industry in Peru. The government had to impose restrictions, including border closures and travel limitations within the country, to prevent the spread of the virus (OXFAM, 2024). As a result, people had to isolate themselves and only go out for essential activities. This led to a decline in the demand and supply of tourism services in Peru and other Latin American countries (Félix et al., 2020). Despite this, cities and the Peruvian government implemented health measures and initiatives to support rural communities (Gestión, 2023). One such initiative was the promotion of regional goods and digital tools to offer virtual touristic visits; it means the facilitation of rural tourism propaganda through Web pages and mobile phone applications that help more potential tourists to access and see the tourist offer, which will increase the intention to visit. This was an effort to restore the losses caused by the pandemic and provide safe experiences for tourists. One of the significant problems faced by rural areas in Peru is the lack of accessible roads, basic services, and adequate lodging. As compared to the capital city of the country, Lima, the percentage of people with access to infrastructure and services is lower. These challenges limit the tourist experience and income from tourism activities for the rural population. Additionally, the lack of infrastructure and basic services affects the quality of life of the local population in these rural areas, leading to increased poverty and reduced development opportunities (Andes Resilient to Climate Change, 2022).

Rural tourism faces several challenges in Peru, including the lack of training and informality in the tourism industry, which can negatively impact the quality, safety, and cost of the travel experience. The COVID-19 pandemic has further exacerbated this issue by increasing the levels of informality. The absence of training and regulatory standards can lead to precarious working conditions, a lack of social security, and decreased tourist satisfaction. In addition, the lack of investment in promotion or marketing initiatives is a limiting factor for rural tourism in Peru. It prevents attracting new visitors, building loyalty among previous visitors, and positioning rural regions as the main tourist destinations in Peru.

The lack of funding for rural areas also affects infrastructure, basic services, quality of life, and the promotion of rural tourism activities. Despite these challenges, institutions such as the Ministry of Foreign Trade and Tourism and PromPeru encourage tourism advertising, marketing, and preservation of cultural and environmental assets in remote regions while providing information to foreign and domestic visitors. However, it also has negative effects on rural people and the environment. The lack of local involvement in decision-making and the unequal distribution of economic activity revenue can lead to social problems in some areas. The lack of local involvement can be explained by little oversight of compliance with rural tourism incentive programmes for women. Likewise, the limited financing by the State at the central level or the poor distribution of economic funds at the local level also hinders the successful development of tourism entrepreneurship programmes for women. Furthermore, rural tourism can raise the cost of regional goods, thereby generating conflicts between tourists and local inhabitants due to a lack of respect for local traditions and customs, as well as increased prices for goods and services.

It is important to note that rural tourism can have negative impacts on the environment and biodiversity if not practised responsibly as environmental impacts generated by ecotourism activity described by Uscuchagua Poma and Yupanqui Sulca (2021)and the negative environmental impacts of tourist activity in the Paccha Falls, Jepelacio District of the Province of Moyobamba (Díaz Noriega, 2021) This can lead to negative consequences such as damage to local flora and fauna, water quality, landscape deterioration, and increased pollution due to visitor traffic and infrastructure development. To ensure that economic benefits are distributed fairly and the environment is preserved in Peruvian rural communities, it is essential to adopt sustainable tourist policies and practices. This can be achieved by collaborating with local communities and introducing policies that encourage sustainable activities. Insecurity is a major problem in Peru, which is expressed through robberies, violence, and human trafficking. This issue affects both citizens and visitors, especially in rural areas where there is a lack of surveillance. Gender-based violence is also prevalent in rural areas or zones far from Peru's main cities. Women tourists may be particularly vulnerable in such areas.

Female Entrepreneurship in Rural Tourism Development

Based on the information provided earlier, we will now showcase three female-led entrepreneurial ventures from Peru's rural tourism sector. We will highlight their key features and explain how they have benefited from the implementation of various social policies, with the support of both private and public institutions.

KURU, from Junín. Anyela Gómez created the socio-environmental company KURU with the aim of fighting extreme poverty in the rural mountains. This initiative innovated in the online production and marketing of organic and natural sorbets, implementing the circular economy to give a new use to a plant previously discarded in the central mountains of Peru. The origin of this venture dates to August 2016, when Anyela Gómez was pursuing a professional degree in Corporate Communication, and in a "Design Thinking" class, the students were asked to find a common solution For an indiscriminate use of plastic straws in the country, given that it is one of the main polluting agents of marine and terrestrial ecosystems. So, Anyela and a group of her classmates undertook research to find a natural product that fulfilled the same functions as a plastic straw without resorting to this synthetic material or other non-renewable resources or high negative impact on the planet. In this way, the university project team – which was initially called "Ayru" – travelled to various parts of Peru, and in the Mantaro Valley they found a high Andean grass known to farmers as "cebadilla", "mala herb", or "bad wheat" that grows wild among potato, wheat, barley, and other crops and is uprooted and burned as it is considered useless and even harmful to agriculture. After consulting with agronomists and other specialists about the potential of this high Andean plant, they verified that with proper processing of this vegetable, sorbets of various diameters could be generated, which makes it possible for there to be special sorbets for each type of drink. Thus, they transform it into natural straws, whose global demand has increased as a replacement for plastic. KURU has benefited more than 20 families, which by including them in the productive chain increased their income by 316%. It should be noted that KURU is made up of 90% women. More details can be seen in https://naturaldrinkingstraws.com

Conclusion

The social policies implemented by the Peruvian government have had a positive impact on the promotion and development of various forms of entrepreneurship in rural tourism. These policies have enabled women to take on new projects and facilitated recognition of their communities at both the national and international levels (MIDAGRI, 2023). The implementation of these social policies has been a significant turning point in the lives of these women and their communities, empowering them to take on greater responsibilities and challenges, and breaking down traditional gender roles. This has led to increased appreciation of female work, greater economic independence, and preservation of their cultural heritage.

References

Abou-Shouk, M. A., Mannaa, M. T., & Elbaz, A. M. (2021). Women's empowerment and tourism development: A cross-country study. *Tourism Management Perspectives*, *37*, 100782. https://doi.org/10.1016/j.tmp.2020.100782

Action Against Hunger. (2022). *Entrepreneurship in 2022: Women, rural environment and environment*. Retrieved 02/02/2024 from https://accioncontraelhambre.org/es/prensa/el-emprendimiento-es-un-elemento-transformador-los-entornos-rurales

Andes Resilient to Climate Change. (2022). *The ancestral knowledge of rural women in the face of climate change as policy enhancers for peasant family farming*. Retrieved 02/02/2024 from https://weadapt.org/knowledge-base/sdc-network-climate-drr-and-environment/andes-resilientes-al-cambio-climatico-resilient-andes/

Cambra-Fierro, J. J., Fuentes-Blasco, M., Huerta-Álvarez, R., & Olavarría-Jaraba, A. (2022). Destination recovery during COVID-19 in an emerging economy: Insights from Perú. *European Research on Management and Business Economics*, *28*(3), 100188. https://doi.org/10.1016/j.iedeen.2021.100188

Carvache-Franco, M., Alvarez-Risco, A., Carvache-Franco, O., Carvache-Franco, W., Estrada-Merino, A., & Villalobos-Alvarez, D. (2022). Perceived value and its influence on satisfaction and loyalty in a coastal city: A study from Lima, Peru. *Journal of Policy Research in Tourism, Leisure and Events*, *14*(2), 115–130. https://doi.org/10.1080/19407963.2021.1883634

Carvache-Franco, M., Carvache-Franco, O., Carvache-Franco, W., Alvarez-Risco, A., & Estrada-Merino, A. (2021). Motivations and segmentation of the demand for coastal cities: A study in Lima, Peru. *International Journal of Tourism Research*, *23*(4), 517–531. https://doi.org/10.1002/jtr.2423

Che, C., Koo, B., Wang, J., Ariza-Montes, A., Vega-Muñoz, A., & Han, H. (2021). Promoting rural tourism in Inner Mongolia: Attributes, satisfaction, and behaviors among sustainable tourists. *International Journal of Environmental Research and Public Health*, *18*(7), 3788. https://doi.org/10.3390/ijerph18073788

Collins-Kreiner, N., & Ram, Y. (2021). National tourism strategies during the Covid-19 pandemic. *Annals of Tourism Research*, *89*, 103076. https://doi.org/10.1016/j.annals.2020.103076

Coroş, M. M., Privitera, D., Păunescu, L. M., Nedelcu, A., Lupu, C., & Ganuşceac, A. (2021). Mărginimea sibiului tells its story: Sustainability, cultural heritage and rural tourism—A supply-side perspective. *Sustainability*, *13*(9), 5309. https://doi.org/10.3390/su13095309

Cusco Peru. (2023). *Conoce Machu Picchu*. Retrieved 02/02/2024 from https://www.cuscoperu.com/es/viajes/machupicchu/

Díaz Noriega, J. C. (2021). Impactos ambientales negativos de la actividad turística en las cataratas de Paccha, Distrito de Jepelacio de la Provincia de Moyobamba.

Elshaer, I., Moustafa, M., Sobaih, A. E., Aliedan, M., & Azazz, A. M. S. (2021). The impact of women's empowerment on sustainable tourism development: Mediating role of tourism involvement. *Tourism Management Perspectives*, *38*, 100815. https://doi.org/10.1016/j.tmp.2021.100815

Félix Mendoza, Á. G., Vera Intriago, D. A., & Cevallos Cevallos, J. K. (2021). Caracterización de corredores gastronómicos como elementos turísticos potenciales en zonas rurales. Caso de estudio: la zona central de Manabí-Ecuador. *Siembra*, *8*(2).

Figueroa-Domecq, C., de Jong, A., & Williams, A. M. (2020). Gender, tourism & entrepreneurship: A critical review. *Annals of Tourism Research*, *84*, 102980. https://doi.org/10.1016/j.annals.2020.102980

Fox, C. F. (2021). Tourism in Peru – Making Machu Picchu: The politics of tourism in twentieth-century Peru. By Mark Rice. Chapel Hill: University of North Carolina Press, 2018, pp. xvi, 233. Abbreviations. Illustrations. Notes. Bibliography. Index. $90.00 cloth; $29.95 paper. *The Americas*, *78*(1), 173–175. https://doi.org/10.1017/tam.2020.121

Gabriel-Campos, E., Werner-Masters, K., Cordova-Buiza, F., & Paucar-Caceres, A. (2021). Community eco-tourism in rural Peru: Resilience and adaptive capacities to the Covid-19 pandemic and climate change. *Journal of Hospitality and Tourism Management, 48*, 416–427. https://doi.org/10.1016/j.jhtm.2021.07.016

Gestión. (2023). *Sustainable tourism: Opportunities and challenges for Peru.* Retrieved 02/02/2024 from https://blogs.gestion.pe/gestiondeservicios/2023/10/turismo-sostenible-oportunidades-y-retos-para-el-peru.html

Government of Peru. (2021). *National strategy for the reactivation of the tourism sector 2022–2025.* Retrieved 02/02/2024 from https://cdn.www.gob.pe/uploads/document/file/3109734/R.%20M.%20N%C2%B0%20138%20-%202022.pdf.pdf

Government of Peru. (2024). *Peru registered more than 2.5 million international tourists in 2023.* Retrieved 02/02/2024 from https://www.gob.pe/institucion/mincetur/noticias/891448-el-peru-registro-mas-de-2-5-millones-de-turistas-internacionales-en-el-2023

Kimbu, A. N., Ngoasong, M. Z., & de Jong, A. (2024). Gender, entrepreneurship and social policy in tourism: Tying the knot. *Journal of Sustainable Tourism*, *32*(3), 421–437. https://doi.org/10.1080/09669582.2024.2309216

Kumar, S., & Shekhar. (2020). Technology and innovation: Changing concept of rural tourism – a systematic review. *Open Geosciences*, *12*(1), 737–752. https://doi.org/10.1515/geo-2020-0183

La Camara. (2020). *Learn about the challenges and opportunities of rural tourism in Peru.* Retrieved 02/02/2024 from https://lacamara.pe/conoce-los-retos-y-oportunidades-del-turismo-rural-en-el-peru/

MIDAGRI. (2022). *MIDAGRI: Entrepreneurship strategy for rural and indigenous women will continue in 2023.* Retrieved 02/02/2024 from https://www.gob.pe/institucion/midagri/noticias/683967-midagri-estrategia-de-emprendimiento-de-la-mujer-rural-e-indigena-continuara-en-el-2023

MIDAGRI. (2023). *Entrepreneurship strategy for rural and indigenous women 2023.* Retrieved 02/02/2024 from https://www.gob.pe/40587-programa-de-compensaciones-para-la-competitividad-estrategia-de-emprendimiento-de-la-mujer-rural-e-indigena-2023

MINCETUR. (2007). *Lineamientos para el desarrollo del turisno rural comunitario en el Perú.* Retrieved 11/04/2024 from https://servilex.pe/documents/ambiente/rm054-2007-mincetur.pdf

Mosedale, S. (2014). Women's empowerment as a development goal: Taking a feminist standpoint. *Journal of International Development*, *26*(8), 1115–1125. https://doi.org/10.1002/jid.3050

National Grographic. (2023). *Why the Nasca lines are among Peru's greatest mysteries.* Retrieved 02/02/2024 from https://www.nationalgeographic.com/history/article/nasca-lines

Ombudsman of Peru. (2023). *Indigenous villages.* Retrieved 02/02/2024 from https://www.defensoria.gob.pe/grupos_de_proteccion/poblacion-indigena/

OXFAM. (2024). *What are the main challenges and demands of rural women in Peru?* Retrieved 02/02/2024 from https://peru.oxfam.org/video-principales-retos-y-demandas-de-las-mujeres-rurales-del-peru

Pérez Gálvez, J. C., López-Guzmán, T., Buiza, F. C., & Medina-Viruel, M. J. (2017). Gastronomy as an element of attraction in a tourist destination: The case of Lima, Peru. *Journal of Ethnic Foods*, *4*(4), 254–261. https://doi.org/10.1016/j.jef.2017.11.002

Priatmoko, S., Kabil, M., Akaak, A., Lakner, Z., Gyuricza, C., & Dávid, L. D. (2023). Understanding the complexity of rural tourism business: Scholarly perspective. *Sustainability*, *15*(2), 1193. https://doi.org/10.3390/su15021193

Rainforest Alliance. (2022). *Indigenous women and entrepreneurs for a more sustainable future*. Retrieved 02/02/2024 from https://www.rainforest-alliance.org/es/en-el-campo/mujeres-indigenas-y-emprendedoras-para-un-futuro-mas-sostenible/

Rosalina, P. D., Dupre, K., & Wang, Y. (2021). Rural tourism: A systematic literature review on definitions and challenges. *Journal of Hospitality and Tourism Management, 47*, 134–149. https://doi.org/10.1016/j.jhtm.2021.03.001

RPP. (2021). *Tourism businesses lost more than s/ 60,000 million due to the pandemic*. Retrieved 02/02/2024 from https://rpp.pe/economia/economia/negocios-de-turismo-en-el-peru-perdieron-60000-millones-de-soles-debido-a-la-pandemia-noticia-1360012?ref=rpp

Ruiz-Ortega, M. J., García-Villaverde, P. M., De La Gala-Velásquez, B., Hurtado-Palomino, A., & Arredondo-Salas, Á. Y. (2021). Innovation capability and pioneering orientation in Peru's cultural heritage tourism destinations: Conflicting environmental effects. *Journal of Hospitality and Tourism Management, 48*, 441–450. https://doi.org/10.1016/j.jhtm.2021.07.012

Sigala, M. (2020). Tourism and COVID-19: Impacts and implications for advancing and resetting industry and research. *Journal of Business Research, 117*, 312–321. https://doi.org/10.1016/j.jbusres.2020.06.015

UNCTAD. (2021). *The global economy could lose more than 4 trillion dollars due to the impact of COVID-19 on Tourism*. Retrieved 02/02/2024 from https://unctad.org/es/news/la-economia-mundial-podria-perder-mas-de-4-billones-de-dolares-por-el-impacto-del-covid-19-en

Uscuchagua Poma, C. E., & Yupanqui Sulca, K. R. (2021). Impactos ambientales generados por la actividad ecoturística y propuesta de mejora en la Reserva Nacional Tambopata, Madre de Dios. https://repositorio.continental.edu.pe/bitstream/20.500.12394/10341/2/IV_FIN_107_TE_Uscuchaga_Yupanqui_2021.pdf

Weyland, F., Colacci, P., Cardoni, A., & Estavillo, C. (2021). Can rural tourism stimulate biodiversity conservation and influence farmer's management decisions? *Journal for Nature Conservation, 64*, 126071. https://doi.org/10.1016/j.jnc.2021.126071

World Bank. (2013). *Peru: A megadiverse country that invests in protected natural areas*. Retrieved 02/02/2024 from https://www.bancomundial.org/es/news/feature/2013/06/06/peru-pais-megadiverso-que-invierte-en-areas-naturales-protegidas-gpan-pronanp

World Bank Blog. (2021). *Global tourism after COVID-19: Three steps for a better recovery of the sector*. Retrieved 02/02/2024 from https://blogs.worldbank.org/es/voices/el-turismo-mundial-despues-de-la-covid-19-tres-pasos-para-una-mejor-recuperacion-del-sector

3 Legitimacy-Seeking in Women's Tourism Entrepreneurship Experience in Turkey

Gizem Kutlu

Introduction

Tourism started to develop in Turkiye in the 1980s, and the 'sustainable tourism' term was first mentioned in the eighth Five-Year Development Plan (2001–2005). With the last 11th Five-Year Development Plan (2019–2023), Turkey aimed to be among the top five most touristed countries in the world in 2023. Although there was a decline in growth rate of tourism industry after COVID-19, Türkiye entered the top five global destinations in 2021 (OECD Tourism Trends and Policies, 2022). This achievement can be considered as the influence of 'Future in Tourism' project started by collaboration between the Turkish Ministry of Culture and Tourism and United Nations to serve the 17 Sustainable Development Goals (United Nations Development Program, 2021). The study of Guzeloglu and Gulc (2021) examined the projects in the 'Future in Tourism' and demonstrated all tourism projects serve one way or another to SDG 5 (gender equality) in Turkey. Thereby, the increasing number of women in tourism sector contributes to the development of sustainable tourism and helps Turkey to take its place internationally among top 5 touristed countries. This shows how the role of gender and women's inclusion plays a crucial role in the development of Turkey's tourism industry and economy. This supports the findings of Alarcón and Cole (2019) claiming that 'there is no sustainability in tourism without gender equality'.

The United Nations World Tourism Organisation's Global Report on Women in Tourism (Ferguson et al., 2019) indicates that women make up 54% of the tourism workforce, while men make up 39%. Moreover, the recent studies from Turkey explicitly claimed that 'tourism is perceived as feminine sector, therefore the number of women employees in tourism sector is higher than men in Turkey' (Maden, 2015; Cicek et al., 2017; Tekeli, 2019; Avci & Gumus, 2022). Nevertheless, Yıldırım et al. (2020) investigated that Turkey has a patriarchal social structure meaning that there are certain gender roles women and men are attributed to do in tourism sector and authority, power, and agency are perceived as masculine. Therefore, women's attempts to undertake any tourism initiatives as entrepreneurs can get a reaction by society.

The discussion table suggest that there is an existing support mechanism in Turkey to increase the number of women workers in tourism sector to achieve

DOI: 10.4324/9781003315858-3

so-called 'gender equality; however, women still work mostly in unqualified, low-paid, or non-paid jobs, while men exist in top management or entrepreneurial positions in tourism sector (Cicek et al., 2017). This shows that there are gender challenges that are specific to 'women entrepreneurs' in tourism industry from Turkey. However, despite these challenges, there is evidence of a growing number of successful women entrepreneurs in tourism and other industries (Bayram et al., 2023). For instance, women-led businesses in sustainable tourism, like eco-friendly resorts and community-based travel experiences foster environmental stewardship, create jobs for local women, and empower communities (Kutlu & Ngoasong, 2023). These achievements highlight the crucial role of women entrepreneurs in driving economy, innovation, and inclusivity (Ozasir Kacar et al., 2023).

The gendered challenges in female entrepreneurship can be overcome through gender-aware policies by creating more inclusive and gender-sensitive environment in Turkey (Sefer, 2020). Following the research agenda in gender, entrepreneurship, and social policy (Ahl & Marlow, 2019; Figueroa-Domecq et al., 2020; Kutlu & Ngoasong, 2023; Kimbu et al., 2024) and the limited research on the intersection of these three concepts from Turkey (Ali & Gursoy, 2020; Sefer, 2020), this book chapter builds an integrative theoretical framework from feminist and institutional theory lenses for understanding the gendered challenges Turkish women entrepreneurs face and how they overcome them through legitimacy-seeking; strategically doing and/or redoing gender.

This chapter contributes to the existing literature on gender, entrepreneurship, and social policy that incorporates critical social role theory provides insights into how individuals assign gender in different ways based on patriarchal context in Turkey. The construction of gender through entrepreneurship and social policy provides opportunities to (re)think how gender might be conceptualised differently. Moreover, the findings of this book chapter offer practical implications for women entrepreneurs and policy implications through social policy suggestions to become more inclusive and sustainable for the greater good.

Literature Review

Women's Tourism Entrepreneurship in Turkey

The tourism industry is a promising field for entrepreneurial ventures in Turkey with its natural beauties, history, seaside resorts, and sport activities. Turkey ranks as the fourth most popular destination globally, boasting a diverse tourism industry that includes various sectors such as hospitality (including guesthouses, hotels, and restaurants), gastronomy, medical tourism, eco-tourism, attractions, and entertainment, as well as transportation and travel information services. However, gender influences are evident in the tourism industry with specific roles traditionally assigned to women and men in their professional pursuits (Cicek et al., 2017). This causes gendered challenges for women aiming to advance their roles within the tourism industry, particularly as women entrepreneurs. Breaking away from traditional roles like cooking and cleaning is a hurdle for Turkish women. With the ongoing sustainable tourism projects, Turkey's sustainable tourism policies emphasise the significance of gender equality by seeking to empower women through financial support and creating new opportunities. However, this chapter uncovers the need for further enhancements with social policies to foster a more inclusive and sustainable tourism sector.

Gender Influences on Geographic Context

Turkish women started tourism entrepreneurship at home (to support their husbands/fathers) with handcrafts, kilims, and jams and sold them to tourists. The legitimacy of women's entrepreneurship in tourism was easy because Turkish women were already doing what they are used to do at home, and therefore they were *doing gender roles* in patriarchal family background in tourism. Legitimacy is defined as a generalised perception that an entity's actions are desirable, proper, or appropriate within a socially constructed system of norms and beliefs (Suchman, 1995, p. 574; Swail & Marlow, 2018). In Turkish culture, carer and hospitality are the role given to women as a housewife. As a result, the family supports a substantial amount of free work done by women in tourism enterprises. This is why Turkish women play a significant role in agriculture, homestays, food and beverage services, and souvenir manufacturing in the tourism industry. Yet, women face many gendered challenges in the institutional context when they want to go beyond these sectors and engage in sustainability in tourism.

Gender Influences on Institutional Context

The institutional environment for women entrepreneurs refers to governmental and non-governmental organisations, policies (e.g. welfare policy, environmental, women's rights, five-year development plans), legislation, finance, and the perception of women in Turkish society (Kutlu & Ngoasong, 2023). Tekeli (2019) highlighted the challenges women entrepreneurs face in tourism sector with their *gender roles in society* (e.g., mothers, wives, carers) and claimed *tourism as 'men dominant sector' limiting decision-making roles of women and access to finance* especially in rural areas (Sefer, 2020). Turkish women are not supported by their families and social circles in their initiatives due to the patriarchal structure of the country (Maden, 2015). These challenges occur in Turkey because the institutional environment is gendered.

Women entrepreneurs are supported with education, entrepreneurship, and consultation, as well as to market the products produced with the support of local governments and non-governmental organisations in Turkey (KOSGEB, 2022). However, even Turkey supports women inclusion in tourism sector for the sustainable development policies, the number of women entrepreneurs in tourism is still less than that of male entrepreneurs (Avci & Gumus, 2022). The studies claim that there are specific challenges to these women in tourism sector because of their gender roles, patriarchal values in society, access to finance and other support mechanisms such as networking, and policy-related challenges (Sefer, 2020).

While women's inclusion with sustainable policy development supports gender equality in Turkey, in many cases it is serving to further entrench gender inequality as mentioned previously women's consistent roles in tourism, low paid jobs, cooking, staying home, and welcoming guests.

Thereby, this book chapter suggests women can overcome these challenges in patriarchal societies by *'doing' and 're-doing' gender as a legitimation strategy* depending on the institutional enablers or challengers they face. In my view, gender roles and relationships are fluid and negotiable. Post-structural feminism goes beyond the concept of women as victims, which suggest that society's dominant codes and behaviours could be reworked, disrupted, challenged, and transformed, so the change is possible (Aitchison, 2005, p. 217; Dowling, 2011).

Drawing on both gender influences and institutional influences on women's tourism entrepreneurship in Turkey, the integrative theoretical framework from feminist and institutional lenses in the following section helps us understand the challenges women entrepreneurs face in tourism industry and how they overcome these challenges through legitimation strategy by doing and redoing gender.

Theoretical Underpinnings of the Study

Gender Role Socialisation

Based on gender role socialisation (Chodorow, 1971), the emerging theoretical model uncovers socially constructed gender influences in women's entrepreneurial decision-making process by entering tourism industry. Women entrepreneurs' socialisation of care ethics is strongly reflected in their triple bottom line goals (i.e., social, environmental, and economic) in their business. Women working in the tourism industry, such as waitresses, housekeepers, and dishwashers in restaurants and guesthouses, have always been recognised as 'innkeeper's wives' or 'restaurant's wives' in Western societies. Besides major tourism facilities in modern Turkey, the care given role leads to successful and cleaning facilities being run by the wife of the owner (Cicek et al., 2017). In this sense, gender role socialisation help explain how Turkish women gain legitimation by 'doing gender' for their entrepreneurship.

Feminist Ethics of Care

When women wanted to go beyond home-related work in tourism, they started to face challenges such as access to market, network, and finance because of their perceived gender roles (Yıldırım et al., 2020). Post-feminist lenses (Gilligan, 1982; Held, 2014; Diaz Garcia & Welter, 2013) help us understand how these women re-construct their feminine and entrepreneurial identities when they interact with their social environment and overcome those challenges by adding value to their femininity in the business realm. Women rethink gender and act in ways that go beyond the normative roles of women, as well as the regulatory structures that penalise women for the roles they play (e.g., caretaker). This construction of identity is also shaped by other factors, such as entrepreneurs' interactions with institutions.

Institutional Theory: Legitimation

Conceptualising institutional factors for women entrepreneurs in a particular country context can help us better understand gender influences as enablers or barriers and provide appropriate social policies for gender equality in both tourism industry and society. In institutional theory (North, 1990), legitimacy plays a central role. Legitimacy refers to whether an agent's actions are desired, appropriate, or proper. Female and male entrepreneurs may be affected in the same way by some regulatory institutions, but others have gendered effects, such as labour market laws granting (un)equal access to employment positions and family policies providing childcare (Welter & Smallbone, 2010).

In this sense, institutional lenses enable us to comprehend institutional challenges that are specific to women entrepreneurs in tourism as well as those gaps in the development of sustainable tourism entrepreneurship (Kolk, 2014). Consequently, the context influences the nature and the type of sustainable tourism entrepreneurship in Turkey, as well as how Turkish female entrepreneurs pursue their business practices in a certain way. Although women entrepreneurship in the field of tourism is supported by the Turkish Government (tangible/loans, providing education), the literature has shown that these supports are not sufficient.

Methodology

This study seeks theory development from in-depth and context-relevant case studies (Miles et al., 2014) of women-owned businesses, where tourism is one of the fastest-growing and among the top 5 touristic destinations as Turkey. However, the number of women entrepreneurs in tourism is still less than that of men in Turkey. On the other hand, the literature shows the contribution of women to the development of tourism industry, economy, and sustainability (Ministry of Culture and Tourism, 2020; Guzeloglu & Gulc, 2021). Gendered challenges caused by patriarchal values in Turkey can be addressed by policy interventions to increase women's participation in sustainable development and to support women's business in tourism (OECD, 2008, p. 11; Sefer, 2020). This book chapter will be useful to social policy in the Turkish context because it examines the influence of gender in tourism business practice.

Moreover, Turkey is chosen as a suitable context for this book chapter because the previous women's tourism entrepreneurship studies from Turkey focus on women repeating their gender roles especially in rural areas through cleaning and cooking at home, yet the chosen cases of tourism businesses in this book chapter include eco-tourism, sustainable handicraft shops for visitors in tourist centres, agribusinesses that supply organic food to the hospitality sector, restaurant chains and, and travel services (Uslu et al., 2015). These reveal women do not always comply their gender roles to gain legitimation in their entrepreneurship by not only doing gender and stay home in tourism industry but also re-doing gender and go beyond what is accepted for women to do in tourism industry.

The research approach involves using the case study method, which includes interviews and references to published or web-based resources. I put specific criteria in the selection of women-led business cases, in line with the three criteria (*sustainable management (organisational), socio-cultural impact, environmental impact)* Turkey implemented for the sustainable tourism industry (Türkiye Sustainable Tourism Industry Criteria, 2022). Relying on the three criteria of Turkey's sustainable tourism industry that this study makes use of, I have identified six women's tourism entrepreneurship cases in total (Table 3.1).

All these SMEs run by women to fulfil the sustainability criteria in the tourism industry. Data collection involves interviews with the women founders as key informants, each lasting about one and a half hour, on their site or via Skype. Additionally, data were collected from publications, websites, and books, as well as interviews with stakeholders in the tourism industry and observers.

The data analyses were based on the theoretical framework developed, which included the application of a coding scheme during the data collection that was guided by content analysis, and thematic analysis approach (Miles et al., 2014). Participants' accounts are keywords or phrases within entire responses captured in

Table 3.1 Informant Profiles of Women Tourism Entrepreneurs

Informant	*Pseudonym*	*Marital Status*	*Sustainable Tourism Sector*	*Location*
W1	Ayse	Widow with children	Organic food seller and restaurant in local bazaar for visitors	Istanbul, Silivri (rural)
W2	Meryem	Married, no child	Eco-tourism	Istanbul, Sile (rural)
W3	Selin	Married, no child	Agri-tourism	Istanbul, Polonezkoy (rural)
W4	Nermin	Married, no child	Sustainable Boutique Hotel; bed and breakfast	Izmir (rural)
W5	Suzan	Married with children	Guesthouse and eco-tourism activities	Bilecik (rural)
W6	Demet	Widow with children	100% Handmade sustainable and traditional accessories shop for visitors	Istanbul (urban)
W7	Azra	Married with children	Medical tourism	Istanbul (rural and urban)

transcripts and fieldnotes (Gioia et al., 2013). The analysed content let us discover how gender influences manifest itself as an entrepreneurial legitimation strategy on women's tourism entrepreneurship when they face gendered challenges.

Findings

Gender Influences on Women's Tourism Entrepreneurship

The three themes emerge in women entrepreneurs' narratives as gender influences: male dominance in tourism, legitimacy-seeking through doing gender and legitimacy-seeking through re-doing gender. The first theme refers to gendered challenges women face because of the male dominance (patriarchy) in the tourism industry as well as the entrepreneurship activity in Turkey. The second theme refers to compliance with gender identity and roles by seeking legitimacy where gender is done (Chodorow, 1971; Gilligan, 1982; Held, 2014), whereas in the third theme (Diaz Garcia & Welter, 2013; Tlaiss & Kauser, 2019; Liu et al., 2022), gender is re-done, and women entrepreneurs challenge the gender differences by legitimacy-seeking in a different way. This study shows that all women engaging doing and/or re-doing gender as their legitimation strategy to start their business as female entrepreneurs and/or to challenge the perceived gender differences because of the patriarchal values. Women's strategic decision-making on doing or re-doing gender to gain legitimation for their entrepreneurial behaviour also depends on the institutional factors (North, 1990) such as support from government and/or family, access to finance, regulations, social perceptions about women, and sustainability in the business realm.

Male Dominance and Gendered Challenges in Tourism Industry

First, the discourse of all women entrepreneurs (W1, W2, W3, W4, W5, W6, W7) demonstrated the male dominance in tourism industry. It is already known that there is a gender segregation between sectors in Turkey (Maden, 2015). Women's acknowledging about patriarchy and the lack of support because of 'being woman/female' still manifest women's doing gender as they are victimising themselves with lack of support because of patriarchy:

> I could not get support for my eco-village project because I am woman. (W1)
>
> Our sector is male-dominated, and men are the top. (W2)
>
> Men do not face the challenges we face as female entrepreneurs. (W4)
>
> Why do you think I started at home, because we are in Turkey... you know how women are perceived? (W5)
>
> My father left me the coastal lands by believing that it is useless, and left his houses and lands in cities and near mountains to my brother… (W6)
>
> In our hotel women are at the top but I see so many hotels usually hire women in Turkey for cleaning, cooking and even making gozleme (pancake) in the extremely hot weather. (W7)

These quotes support the claims from the literature such as when discussing gender in tourism and entrepreneurship, men are portrayed as highly dominant role models (Figueroa-Domecq et al., 2020), and due to social and cultural restrictions about women's roles in communities (which can be seen as 'stay at home if you are women' perspective in Turkey), many Islamic and patriarchal countries, including Middle Eastern countries, practise gender inequality in the tourism industry (Hutchings et al., 2020).

Second, access to finance emerged as one of the biggest gendered challenges women entrepreneurs faces in the tourism industry because of their socially constructed gender. The narratives of W5, W2, W1, W3, and W4 highlighted the bias against women from investors and bank loaners and even from financial support mechanisms of government.

> I could not get financial support from the local authorities until I make our village the most visited one by tourists which was on the media, while my male friend in the tourism industry received lots of support without even starting [...] (W5)
>
> The financial support for tourism entrepreneurs is not applicable for most women entrepreneurs who are in the medical tourism, eco-tourism or other alternative tourism industries. (W2)
>
> The credit loaner asked me if I am married and have kids, claiming that it might be difficult to be a woman entrepreneur in the tourism industry with our responsibilities at home. (W1)
>
> I needed my husband to receive funding from bank. (W3)
>
> There is financial support for women entrepreneurs but the criteria is so challenging. (W4)

While the narrative of these women shows the challenges in access to finance, they also highlight the role of media-driven conditions for local authority support (W5), limitations on financial assistance in specific tourism sub-sectors (W2), and biases against women's personal life, including marital status and family responsibilities. The narratives also highlight instances of women depending on male relatives for financial transactions and facing stringent criteria for available support. Overall, these stories reveal the need for addressing gender biases and promoting more inclusive financial support structures to empower women in diverse segments of the tourism industry. For example, revealing women's success in tourism on media (TV and/or social media) from rural areas can increase access to finance while reducing biases against women entrepreneurs.

The third gendered challenge women entrepreneurs face in was the lack of trust on their decision-making when wanted to expand or grow their businesses sustainably.

> Everybody laughed at me when I said them, I can do eco-tourism. (W2)
>
> We run this business with my husband and when I first told him about being environmentally friendly hostel, he thought that it is silly idea but

> then the government enforced some environmental regulations for those in the tourism industry. (W5)
>
> When I first said I wanted to sell my accessories in the bazaar and providing sustainable fashion workshops to tourists, my family insisted me to continue from home. (W6)

The narratives of other women entrepreneurs supported these quotes when they talked about how their decisions are criticised by others either at home or in the business realm. The quotes also showed that it was more challenging for women entrepreneurs to go beyond home-based or/and profit-oriented businesses in the tourism industry. However, the second and third theme revealed how women gain legitimacy for their entrepreneurial decisions and activities through doing or re-doing gender strategically.

Legitimacy Seeking through Doing Gender

For instance, W1, W2, W5 and W6 first started their businesses at home because they felt more comfortable to gain legitimation for the male-dominated entrepreneurship activity in Turkey from both their families and society. While entrepreneurship is considered as male-related activity in Turkey, tourism sector as including feminine features such as hospitality, cooking, producing, and caring guests opens the doors for women to become entrepreneurs.

> The expectation was women should work at home in tourism and help their husbands who are the owners when I started in 2010. (W1)
>
> My family would not support me at the beginning if I told them I will sell these products at the bazaar. (W2)
>
> I am a mother with responsibilities of caring so, when me and my husband started to provide our house as guesthouse to tourist, I felt like I am repeating same thing at home and work. (W5)
>
> I was a married woman when I first started at home and I had to do a lot of housework, take care of my husband and children, so it seemed impossible to do something else (refers to tourism entrepreneurship). I was doing handmade materials at home and my relatives were selling them in their shop for tourists. (W6)

Constructing their identities as women, female entrepreneurs, or married women, these women strategically do gender when they first wanted to start their own business and earn money. They strategically showed that ‘they are not doing something forbidden or shameful for women’. As Tucker (2007) claimed Turkish women were not allowed to work outside the home in tourism industry and it is still the case in some rural areas (Tekeli, 2019). However, the narratives showed how women found a way to start their business through doing gender to gain legitimation and contribute to the family income as well as the Turkey’s economy (Cicek et al., 2017).

Legitimacy Seeking through Re-Doing Gender

Interestingly, this study showed women entrepreneurs do not always 'do gender' and comply with their expected gender roles to gain legitimation for their entrepreneurship but they re-do gender when they face challenges and go beyond the expectations from them. Aligning with narratives of Turkish women from the previous section, their 'carer' role allowed them to be one step further than men when it comes to sustainable entrepreneurship (Outsios & Farooqi, 2017). This means women challenged gender differences by being sustainable entrepreneurs, which help them legitimatise their entrepreneurship because sustainability requires 'ethics of care' (Hechavarría, 2016). In another word, sustainable decision-making allowed women to enjoy their femininity in the business realm while also making profit in Turkey.

For instance, W1 and W2 challenge gender differences and show how they act what is perceived as masculine (e.g., making money, being breadwinner) through their ethical business. This shows the importance of 'feminist ethics of care' because this demonstrates that it is possible to have ethical values and make profit with the same way men do.

> I am a female entrepreneur, but I took many risks as men do [...]
>
> (W1)

> I have the agency to develop and grow my business, I make a lot of profit each year based on my ethical decisions.
>
> (W2)

Similarly, W3 AND W4 re-do gender by adding value to being feminine in the business realm. It is manifested through relational and supportive partnerships, which help them gain access to networking, finance, and mentoring support. Moreover, women value their femininity because it helps them to gain legitimation for sustainable entrepreneurship and enable them to grow their business in size and scale. For instance, W3 said:

> I struggled to receive financial support from government but being part of Women's Entrepreneurship Association where there is always networking opportunities with other women. Building relationships as a female entrepreneur helped me to find an angel investor for my eco-buildings in the village.

Similarly, W4 said:

> Although it would be difficult to do business in a traditional way in Turkey's patriarchal environment (she refers to profit-oriented male dominated way), [...] being a female sustainable entrepreneur helps me to grow my business because women always care others and sustainable entrepreneurship is something you do not only consider profit like men but also people, planet and even your culture.

In this sense, W7 claimed:

> I think when it comes to sustainable entrepreneurship, women are luckier than men in Turkey, imagine as women we have to think about home, husband, children, cooking, cleaning so our brain works like this, so caring others in business something we are very strong at it.

These quotes uncover how women overcame the challenges of being women in a male-dominated society by adding value to their femininity such as collaborating, being relational, and caring others. For instance, W5 who is the founder of the guesthouse with eco-tourism activities is chosen as the Turkey's Chanagemaker by Sabanci. Turkey's Changemakers Program supports sustainable entrepreneurs, activists, and change-actors that produce innovative solutions to social problems and makes them visible.

> Changemakers Network gave me the chance to cooperate with other entrepreneurs and attend various workshops […] There was another changemaker and their business provide market opportunities for fair producers and consumers. This helped me to learn producer economy while creating a market for my organic products.
>
> (W5)

> Being relational and cooperative female helped me a lot to open my touristic shop. There are non-profit organisations that empower low-income women artisans, so when my husband died, nobody believed I will be an entrepreneur, but the women artisan workshops I attended helped me to advertise my handcrafts on social media and then open my shop.
>
> (W6)

These quotes showed how women overcame the patriarchal values and beliefs against their perceived gender roles by re-doing gender through practising what is perceived masculine and adding value to their femininity as being relational or collaborative. Last but not least, supporting other women in and with their business appeared as an important strategy to overcome patriarchy in business.

> I convinced other women in the village to earn and enjoy by doing eco-tourism, so they opened their home to visitors as well.
>
> (W6)

> Women empower women, and I believe by working women who cares the nature and environment we increased the eco-tourism activity sales.
>
> (W7)

'Collaborating other women in the industry' and 'hiring women' and/or 'supporting women through training/seminars/workshops' and how this developed women's business in both size and scale emerged in the quotes of other women entrepreneurs (W1–W2–W3–W4–W5) as well.

The findings suggest that women do gender strategically when there is no other option to step into entrepreneurship; however, when they want to develop their business activities, doing gender does not help because all of them acknowledged the patriarchal environment especially in tourism sector. The reason why women easily start their business in tourism was because it is perceived as a feminine sector and help these women to legitimise what they are doing as a female. However, the sector was still male dominated in Turkey (Cicek et al., 2017). That was the reason why most women started their business at home with 'women should stay at home' perception in the society and actually 'do gender' for legitimation.

Yet, when women realised that they cannot go beyond staying home, challenges in access to finance such as 'needing my husband to receive funding' or 'could not get support because I am women' as stated in this study, women started to 're-do' gender with different ways. It is found that defining themselves as a sustainable entrepreneur with 'feminist ethics of care' and adding value to being feminine and ethical values as well as supporting other women, being relational and cooperative with non-governmental institutions help these women. None of the women mention about the governmental support although Turkey aims to decrease gender gap in tourism industry. This study reveals the role of women in sustainability with their social, cultural, economic, and environmental impact in their business and the tourism industry.

Discussion

The narratives of women entrepreneurs in the Turkish tourism industry highlight three significant gender-related challenges: the prevalence of male dominance (patriarchy), access to finance, and a lack of trust in women's decision-making for sustainable tourism business goals, and most of these gendered challenges occur in the institutional environment. In this sense, the narratives also shed light on how media could be effective for local authority support for women, financial support for women in specific tourism sub-sectors (e.g. eco-tourism, medical tourism), and gender aware discourse against women's personal lives in policies. For example, while the findings of Ditta-Apichai et al. (2023) uncover the importance of social media (Facebook), Santafe-Troncoso and Tanguila-Andy (2023) reveal the role of informal online communities and other media platforms to be effective in increasing women's entrepreneurial behaviour in specific regions. The formal media (TVs, newspapers) seem to be effective for increasing women's visibility in tourism industry from Turkey.

Yet, when Turkish women face these gendered challenges and there is no support in the institutional environment, women develop entrepreneurial strategies with their own decision-making. They seek legitimacy for their entrepreneurship through compliance or challenging which manifests itself as doing and/or re-doing gender. The legitimacy strategy involves complying with, or doing, gender bias to legitimise themselves (Swail & Marlow, 2018), while challenging gender biases entails re-doing gender to legitimise themselves as sustainable entrepreneurs. This study not only contributes to the existing debate on gendering sustainability in sustainable tourism by recognising how sustainability legitimises change and continuity (for women) in relation to tourism (Tucker, 2022; Kimbu et al., 2024) but also offers an overview of the sustainable business practices used by women entrepreneurs in search of legitimacy and change.

Integrating sustainability by practising the feminist ethics of care (Gilligan, 1982; Fors & Lennerfors, 2019) is key for women to gain legitimation while re-doing gender and adding value to femininity in the business realm. These findings can be useful for women entrepreneurs who want to practise 'sustainability' since the literature emphasises the importance of ethics of care (care, support, being relational and collaborative) for the legitimacy of sustainable entrepreneurship (Elkington, 1998). External networks are also found critical for helping women entrepreneurs to re-do gender and gain legitimacy for their entrepreneurship. Thereby, findings suggest that being part of networks, including governmental, multinational corporations, and non-governmental organisations (e.g., women's cooperatives), is critical as these networks enable women to gather information, resources, access to finance, and advice as suggested by studies from other contexts such as Kazakhstan and Mexico (Filimonau et al., 2024; Khoo et al., 2024).

In terms of policy implications, Turkey does not have specific policy that is providing funded nursery school opportunities for women entrepreneurs, and this could be very helpful in Turkey and in other patriarchal societies. Supported social policies relevant to women highlighted in other context for women entrepreneurs such as access to finance, infrastructure, resources, and networking (Ahl & Nelson, 2015; Ngoasong & Kimbu, 2019). Finally training opportunities in entrepreneurship, access to finance for women who wants to go beyond 'cooking, cleaning, or producing touristic goods at home in the tourism industry especially for those in eco-tourism, agrotourism, and medical tourism should be highlighted in the sustainable tourism policies to be more inclusive and follow grassroot approach (Kimbu et al., 2024).

Conclusion

This chapter reveals the experiences of women entrepreneurs from a patriarchal tourism destination, Turkey. The narratives of women show the challenges they faced in their entrepreneurial journey; however, doing gender and re-doing gender strategically for gaining legitimation help them through sustainable business practices. Women entrepreneurs include sustainable management and environmental and sociocultural impact on their business by contributing to the circular economy in Turkey. These women uncover the importance of femininity in the business realm and how being relational, collaborative, and supportive help them to become sustainable entrepreneurs. The theoretical framework contributes to gender, sustainable entrepreneurship, tourism, and social policy literature by integrating different perspectives from feminist and institutional theories and revealing how gender is used for legitimation by women entrepreneurs depending on other contextual, spatial, and institutional factors.

Although there is an ongoing support for women entrepreneurs in Turkey usually through non-governmental institutions, the findings of this study shows that there is still long way to go to achieve gender equality in entrepreneurship. Therefore, this study suggested policy implications that would be helpful to increase the place and role of women entrepreneurs in tourism industry. However, the findings are limited to seven women-owned businesses. Future studies might involve more female entrepreneurs from different patriarchal regions to reveal experiences of these women who face gendered challenges and develop the practical and policy implications of this study.

References

Ahl, H., & Marlow, S. (2019). Exploring the false promise of entrepreneurship through a postfeminist critique of the enterprise policy discourse in Sweden and the UK. *Human Relations*, *74*(1), 41–68.

Ahl, H., & Nelson, T. (2015). How policy positions women entrepreneurs: A comparative analysis of state discourse in Sweden and the United States. *Journal of Business Venturing*, *30*(2), 273–291. https://doi.org/10.1016/j.jbusvent.2014.08.002

Aitchison, C. C. (2005). Feminist and gender perspectives in tourism studies: The social-cultural nexus of critical and cultural theories. *Tourist Studies*, *5*(3), 207–224. https://doi.org/10.1177/1468797605070330

Alarcón, D. M., & Cole, S. (2019). No sustainability for tourism without gender equality. *Journal of Sustainable Tourism*, *27*(7), 903–919.

Ali, F., & Gursoy, D. (2020). Women entrepreneurship in tourism: A systematic literature review. *Journal of Hospitality and Tourism Management*, *44*, 64–77.

Avci, N., & Gumus, S. (2022). The motivations of women entrepreneurs in tourism industry. *Istanbul Business Research*, *51*(1), 279–290.

Bayram, G. E., Bayar, S. B., & Bayram, A. T. (2023). The role of women entrepreneurs on family businesses in Turkey. In Valeri, M. (Ed.), *Family businesses in tourism and hospitality. Tourism, Hospitality & Event Management* (pp. 109–122). Cham: Springer.

Chodorow, N. (1971). Being and doing: a cross-cultural examination of the socialization of males and females. In Chodorow, N. J. (Ed.) *Feminism and psychoanalytic theory* (pp. 23–44). New Haven: Yale and London: Polity Press.

Cicek, D., Zencir, E., & Kozak, N. (2017). Women in Turkish tourism. *Journal of Hospitality and Tourism Management*, *31*, 228–234.

Diaz Garcia, M. C. D., & Welter, F. (2013). Gender identities and practices: Interpreting women entrepreneurs' narratives. *International Small Business Journal: Researching Entrepreneurship*, *31*(4), 384–404. https://doi.org/10.1177/0266242611422829

Ditta-Apichai, M., Gretzel, U., & Kattiyapornpong, U. (2023). Platform empowerment: Facebook's role in facilitating female micro-entrepreneurship in tourism. *Journal of Sustainable Tourism*, *32*(3). https://doi.org/10.1080/09669582.2023.2215479

Dowling, S. J. (2011). *Constructing identity identity construction* [Thesis]. Georgia State University. https://doi.org/10.57709/2093433

Elkington, J. (1998). *Cannibals with forks: The triple bottom line of 21st century business*. Oxford: Capstone.

Ferguson, L., Boonabaana, B., Khoo-Lattimore, C., & Alarcon, D. M. (2019). *Global report on women in tourism* (2nd ed.). Washington, DC: World Bank Group.

Figueroa-Domecq, C., de Jong, A., & Williams, A. M. (2020). Gender, tourism and entrepreneurship: A critical review. *Annals of Tourism Research*. https://doi.org/10.1016/j.annals.2020.102980

Filimonau, V., Matyakubov, U., Matniyozov, M., Shaken, A., & Mika, M. (2024). Women entrepreneurs in tourism in a time of a life event crisis. *Journal of Sustainable Tourism, 32*(3), 457–479 https://doi.org/10.1080/09669582.2022.2091142

Fors, P., & Lennerfors, T. T. (2019). The individual-care nexus: A theory of entrepreneurial care for sustainable entrepreneurship. *Sustainability, 11*(18), 4904. https://doi.org/10.3390/su11184904

Gilligan, C. (1982). *In a different voice*. Cambridge, MA: Harvard University Press.

Gioia, D. A., Corley, K. G., & Hamilton, A. L. (2013). Seeking qualitative rigor in inductive research: notes on the gioia methodology. *Organizational Research Methods, 16*(1), 15–31.

Guzeloglu, F. T., & Gulc, A. (2021). Sustainable tourism projects in Turkey. *Academy of Management Learning and Education, 5*(1), 55–73.

Hechavarría, D. M. (2016). Mother nature's son? The impact of gender socialization and culture on environmental venturing. *International Journal of Gender and Entrepreneurship, 8*(2), 137–172. https://doi.org/10.1108/IJGE-10-2015-0038

Held, V. (2014). The ethics of care as normative guidance: Comment on Gilligan. *Journal of Social Philosophy, 45*(1), 107–115. https://doi.org/10.1111/josp.12051

Hutchings, K., Moyle, C.-L., Chai, A., Garofano, N., & Moore, S. (2020). Segregation of women in tourism employment in the APEC region. *Tourism Management Perspectives, 34*, 100655. https://doi.org/10.1016/j.tmp.2020.100655

Khoo, C., Yang, E. C. L., Tan, R. Y. Y., Alonso-Vazquez, M., Ricaurte-Quijano, C., Pécot, M., & Barahona-Canales, D. (2024). Opportunities and challenges of digital competencies for women tourism entrepreneurs in Latin America: A gendered perspective. *Journal of Sustainable Tourism, 32*(3), 519–539 https://doi.org/10.1080/09669582.2023.2189622

Kimbu, A. N., Ngoasong, M. Z., Adeola, O., & Afenyo-Agbe, E. (2019). Collaborative networks for sustainable human capital management in women's tourism entrepreneurship: The role of tourism policy. *Tourism Planning & Development, 16*(2), 161–178. https://doi.org/10.1080/21568316.2018.1556329

Kimbu, A. N., Ngoasong, M. Z. & Anna, J. (2024). Gender, entrepreneurship and social policy in tourism: Tying the knot. *Journal of Sustainable Tourism, 32*, 421–437. https://doi.org/10.1080/09669582.2024.2309216

Kolk, A. (2014). Linking subsistence activities to global marketing systems: The role of institutions. *Journal of Macromarketing, 34*(2), 186–198.

KOSGEB (2022) SME Definition in Turkey [Online]. Available at Definitions and Regulations - KOSGEB - Small and Medium Enterprises Development Organization. Accessed on 11 August 2022. https://en.kosgeb.gov.tr/site/tr/genel/detay/5667/definitions-and-regulations.

Kutlu, G., & Ngoasong, M. Z. (2023). A framework for gender influences on sustainable business models in women's tourism entrepreneurship: Doing and re-doing gender. *Journal of Sustainable Tourism*. https://doi.org/10.1080/09669582.2023.2201878

Liu, A., Mimi Li, B., & Monica, W. C. (2020). Performing femininity: Women at the top (doing and undoing gender). *Tourism Management, 80*(2020). https://doi.org/10.1016/j.tourman.2020.104130

Maden, C. (2015). A gendered lens on entrepreneurship: Women entrepreneurship in Turkey. *Gender in Management: An International Journal, 30*(4), 312–331.

Miles, M. B., Huberman, A. M., & Saldaña, J. (2014). *Qualitative data analysis: A methods sourcebook and the coding manual for qualitative researchers.* Thousand Oaks, CA: SAGE.

Ministry of Culture and Tourism. (2019). *Supporting women's entrepreneurship in tourism*. Government of Turkey.

Ngoasong, M. Z., & Kimbu, A. N. (2019). Why hurry? The slow process of growth in women-owned businesses in resource-scarce contexts. *Journal of Small Business Management*, *57*(1), 40–58. https://doi.org/10.1111/jsbm.12493

North, D. (1990). *Institutions, institutional change and economic performance*. Cambridge: Cambridge University Press.

OECD. (2008). *Gender and sustainable development: Maximising the economic, social and environmental role of women*. Paris: OECD Publishing [Online]. https://www.oecd-ilibrary.org/environment/gender-and-sustainable-development_9789264049901-en

OECD. (2022). *OECD tourism trends and policies 2022*. Paris: OECD Publishing [Online]. Available at OECD Tourism Trends and Policies 2022 (oecd-ilibrary.org).

Outsios, G., & Farooqi, S. A. (2017). Gender in sustainable entrepreneurship: Evidence from the UK. *Gender in Management: An International Journal*, *32*(3), 183–202.

Ozasir Kacar, S., Essers, C., & Benschop, Y. (2023). A contextual analysis of entrepreneurial identity and experience: Women entrepreneurs in Turkey. *Entrepreneurship & Regional Development*, *35*(5–6), 460–481.

Santafe-Troncoso, V., & Tanguila-Andy, A. (2023). Indigenous women's approaches to tourism planning: Lessons from Ecuador. *Journal of Sustainable Tourism*, *32*(3). https://doi.org/10.1080/09669582.2023.2247580

Sefer, B. (2020). A gender- and class-sensitive explanatory model for rural women entrepreneurship in Turkey. *International Journal of Gender and Entrepreneurship*, *12*(2), 191–210.

Suchman, M. C. (1995). Managing legitimacy: Strategic and institutional approaches. *The Academy of Management Review*, *20*(3), 571–610.

Swail, J., & Marlow, S. (2018). Embrace the masculine; attenuate the feminine - gender, identity work and entrepreneurial legitimation in the nascent context. *Entrepreneurship and Regional Development*, *30*(1–2), 256–282.

Tekeli, H. N. (2019). Women's employment in tourism sector in Turkey, issues faced, and the effect of glass ceiling syndrome on women workforce in tourism. *Turkish Studies Social Sciences*, *14*(2), 179–191.

Tlaiss, H. A., & Kauser, S. (2019). Entrepreneurial leadership, patriarchy, gender, and identity in the Arab world: Lebanon in focus. *Journal of Small Business Management*, *57*(2), 517–537.

Tucker, H. (2007). Undoing shame: Tourism and women's work in Turkey. *Journal of Tourism & Cultural Change*, *5*(2), 87–105. https://doi.org/10.2167/jtcc089.0

Tucker, H. (2022). Gendering sustainability's contradictions: Between change and continuity. *Journal of Sustainable Tourism*, *30*(7), 1500–1517.

Turkey Eight Development Plan. (2001–2005). *The Grand National Assembly of Turkey* [Online]. Long-Term Strategy and Eight Five Year Development Plan 2001–2005 (sbb.gov.tr)

Turkey Eleventh Development Plan. (2019–2023). *The Grand National Assembly of Turkey* [Online]. Available at Eleventh Development Plan 2019–2023 (sbb.gov.tr).

Turkiye Sustainable Tourism Industry Criteria. (2022). *Turkiye Tourism Promotion and Development Agency* [Online]. Türkiye Sustainable Tourism Industry Criteria | TGA. https://cdn.tga.net.tr/tga-prod/document_pool/2024/2/7/EN-Surdurulebilir-Turizm-Endustri-Kriterler_Vol-1-19-Mayis-2022_v2_(2)-dd2104c0-5f29-49de-969c-7e33c9ff968c.pdf

UNDP Turkey. (2021). *Supporting women's entrepreneurship in tourism in Turkey*. New York: United Nations Development Programme.

Uslu, Y. D., Hancıoğlu, Y., & Demir, E. (2015). Applicability to green entrepreneurship in Turkey: A situation analysis. *Procedia - Social and Behavioral Sciences, 195*, 1238–1245. https://doi.org/10.1016/j.sbspro.2015.06.266

Welter, F., & Smallbone, D. (2010). The embeddedness of women's entrepreneurship in a transition context. In Brush, C. G., et al. (Eds.), *Women entrepreneurs and the global environment for growth: A research perspective* (pp. 96–117). Cheltenham: Edward Elgar.

Yıldırım, G., Tarınç, A., & Kılınç, C. C. (2020). Women entrepreneurship in tourism: The case of Turkey. *Journal of Tourism and Gastronomy Studies, 8*(4), 2462–2477.

4 Hotel House-Keepreneurship

Case Study of "Las Kellys" in Spain

Almudena Otegui Carles, José A. Fraiz-Brea and Noelia Araújo Vila

Introduction

The Context of "Las Kellys" in Spain

Spanish hotel housekeepers created in 2016 an association known as Las Kellys, with the aim of claiming for their rights. The name Las Kellys (feminine and plural name) comes from a pun "Las que limpian" (Who cleans). Those involved in the association initially started getting together through discussion groups, public events, and social media in 2014 (Cañada, 2015). Throughout 2015, they formed a series of social media groups, and at the beginning of 2016, they were formally inscribed in the register of Spanish associations, the first association of hotel housekeepers: Las Kellys (Kellys, 2015). This first association was established at the national level, but then more associations began to be created at regional and provincial levels. According to the Directory of Association Names of the Spanish Ministry of the Interior, as of April 2023, there were 18 associations of Las Kellys at national, regional, and provincial levels (Ministerio del Interior, 2023).

The first action of Las Kellys was in Benidorm on August 16, 2016, whilst October 12, 2016, in Barcelona was their official presentation. To increase their organizational power without losing their individual representation, some of the associations have decided to join a federation, which is a grouping of associations. They have also formed an NGO platform called Pecapis (State Platform of Hotel Housekeepers), which was created in April 2020 (Navarro, 2020), to help hotel housekeepers that had a temporary contract in the hotels where they were dismissed prior to the declaration of the state of alarm in Spain (Sevilla, 2020) because of the pandemic. Under the law, associations are not official trade unions, and this leaves them completely removed from any possible legal representation and ensures that they may not participate in the sectoral agreement negotiations (Moral-Martin, 2020). For this reason, in 2018, Las Kellys Cataluña decided to become a trade union.

The demands of these associations are, among others, the right to early retirement; linking the category of hotels to the quality of the work they generate; the recognition of their occupational diseases; the increase in labour inspections; the implementation of ergonomic studies; and compliance with the labour collective

DOI: 10.4324/9781003315858-4

agreements (Cañada, 2015; Delgado, 2018; Otegui Carles et al., 2023). At regional level Las Kellys have made progress in different regions of Spain, but they have never achieved their main requirement, the end of outsourcing by the amendment of the article 42.1 of the 2015 Spanish Workers' Statute (Kellys, 2015), which represents precariousness due to loss of rights. In Catalonia, though, they have managed to encourage the Catalan Parliament to approve a Seal of Quality and Fair work for the tourism and hospitality sector, which must include (Catalunya, 2018): The application and compliance with the interprovincial collective agreement of the sector of the hospitality and tourism industry of Catalonia; The non-outsourcing of central services in the hotel business, such as hotel housekeepers, to third parties; Compliance with occupational risk prevention laws, especially regarding work overload; Absence of a history of serious work infractions; Equal pay for men and women for equal work; The hiring of people who are part of vulnerable groups; and the existence of measures related to the promotion of stable and quality employment.

However, as of November 2022 the stamp had not been implemented. This is what led to Las Kellys Cataluña to enact a form of entrepreneurship by creating a hotel booking platform, financed through crowdfunding. This booking platform intends to sell hotel rooms from hotels where "We can assure, with total certainty, that the rights of the workers are respected and there is no exploitation" (Goteo, 2021). The enterprise was called "Yo reservo con Las Kellys" (I book with Las Kellys). Through crowdfunding they initially sought a minimum of €60,000 yet obtained more than €90,000. To contribute there were five options, from €15 to €500, with different rewards, from "get a clean word" to "be invited to the great opening party of the platform that will take part in Barcelona, and to include the name of the contributor in the acknowledgments of the website of the platform". Through the Las Kellys Cataluña platform, travellers could book their rooms in hotels that meet the requirements of the Seal of Quality and Fair Work (Goteo, 2021). The booking platform will start operating in 2023, which would be a milestone in Spain since it would be the first entrepreneurial activity conducted by a legally established association of female hotel housekeepers.

It was not until the 1990s that the scientific literature urged women to become entrepreneurial subjects (Serrano-Pascual & Carretero-García, 2022). As stated by Freund and Hernandez-Maskivker (2021, p. 5) 20 years after: "Deeper studies within the field of tourism on women's associations and networks are needed to achieve a greater understanding of the sector on this issue". Therefore, with the objective to increase the limited research on women's associations, the aim of this chapter is to find out how, within the socio-political context of the existing associations of hotel housekeepers in Spain, women hotel housekeepers could improve, through entrepreneurship in the form of a hotel booking platform their rights and working conditions.

To achieve this aim, first a literature review is presented to expose a background that informs the various components being researched. The method is next introduced, detailing how semi-structured in-depth interviews were conducted to the spokespersons of different Las Kellys associations. After that, results and findings are presented to answer the questions raised in relation to the research aim, before turning to outline the conclusion, which includes the main limitations found, and potential future lines of research. The interest of this study lies not only in deepening this new female entrepreneurship but in establishing the connection it will have with the struggle for workers' rights because its success or failure can point out the way for other vulnerable groups in the tourism sector to fight for their rights, both in Spain and in a hypothetical dissemination to other countries.

Literature Review

Hotel Housekeepers Work Conditions

Women frequently carry out the lowest status work in hospitality (Mooney & Ryan, 2009), whilst they experience the greatest health inequalities among workers in the industry (Rosemberg, 2020). The job of hotel housekeepers is associated with poor working conditions, time pressure, low job control, and job insecurity (Nimri et al., 2020).

Nine out of ten psychosocial risk factors identified by the World Health Organization are suffered by hotel housekeepers (Hsieh et al., 2016). Hotel housekeepers are one of the sectors with the highest rates of accidents at work and occupational diseases (Cañada, 2015). Precariousness and outsourcing are two of the most alarming problems as they lead to worse working conditions, fewer labour rights, and lower wages (Cañada, 2015). They suffer physical, chemical and biological, and psychological risks (Hsieh et al., 2013) that lead to premature osteoarthritis, disc protrusions, carpal tunnel syndrome, painful shoulder, tendinitis, stress, anxiety, or depression (Otegui Carles et al., 2023), making it necessary for such workers to take acetaminophen, ibuprofen, dexketoprofen, diclofenac, nolotil, valerian, alprazolam, hypnotics like diazepam, and, when these are not enough, opioids (Cañada, 2015). During the pandemic there has been a worsening of working conditions, an increase in unethical work practices, extra pressure, higher workload, and job insecurity, all of which in turn has led to an increase in their stress (Otegui Carles et al., 2023).

As a consequence, the associations of hotel housekeepers in Spain claim for the recognition of diseases directly related to the motor apparatus and skeletal muscle as occupational diseases, recognition of hard work for early retirement, decent pensions, greater detection of exploitative practices, and prevention of fraudulent behaviour through ensuring inspections are more effective as well as an end to outsourcing which is understood to lead to a worsening of the work conditions (Alcalde-González et al., 2021; Cañada, 2015; Delgado, 2018; Otegui Carles et al., 2023).

House-Keepreneurship

Building on the tourism entrepreneurship literature, this chapter introduces the concept of hotel house-keepreneurship to investigate the claiming of worker rights by female hotel housekeeper entrepreneurs. There are different motivations to be an entrepreneur, such as the dissatisfaction in the previous work place because of demanding and inflexible work environments (Carvalho et al., 2018), segregation to job occupations or positions (Costa et al., 2016; Russen et al., 2021), work–life balance (Ribes-Giner et al., 2018), personal development and fulfilment, difficulties in labour market integration (Figueroa-Domecq et al., 2020), and recognition (Noguera et al., 2015).

Entrepreneurial behaviours are influenced by factors such as age, gender, management skills, psychological, sociocultural norms, and gender role stereotypes (Kourtesopoulou & Chatzigianni, 2021); government financial support, administrative procedures, technological development, and economic incentives (Butkouskaya et al., 2020); education, family context, or income level (Noguera et al., 2015). Women entrepreneurs face socio-cultural, political, and institutional constraints (Ribeiro et al., 2021) as well as barriers to access resources such as mentors, investors, and advisors (Neumeyer et al., 2019) having to pursue personal and/or family funding and informal microfinance (Figueroa-Domecq et al., 2022).

Crucially, belonging to networks can help women entrepreneurs overcome constraints and barriers (Kimbu et al., 2019), with these networks being critical to supporting and promoting entrepreneurship (Neumeyer et al., 2019) and enhancing women entrepreneurs' capacity and capabilities (Ribeiro et al., 2021). Women face difficulties to pursue business and political ties because of stereotypes and because they have less time due to normative familial caring roles, that is, they are often expected to combine with their entrepreneur role (Ribeiro et al., 2021). For these reasons, women have been found to limit their ties to networks of women entrepreneurs (Ribeiro et al., 2021) including, for example, women's associations.

Nevertheless, women's associations provide support on a local, national, and international level transforming the economic, social, and political context of societies (Freund & Hernandez-Maskivker, 2021). They help women to solve problems, alleviating women of daily concerns, improving their lives, and empowering them (Freund & Hernandez-Maskivker, 2021). Despite previous studies pointing to the fact that precarious workers usually employ subtle tactics rather than formal collective action (Alcalde-González et al., 2021), in Spain hotel housekeepers have confronted precariousness through formal associations that have helped female hotel housekeepers and have created collective resistance in the struggle for their rights. These associations have built a dignified occupational identity and have become a socio-political agent in the Spanish public arena (Alcalde-González et al., 2021).

Method

To understand how the socio-political context has led hotel housekeepers to undertake a form of entrepreneurial through the creation of an online hotel booking platform, two qualitative methods have been used: semi-structured in-depth interviews, and secondary data analysis of official documents, legislation, websites, journals, and newspapers. To conduct the interviews, two scripts were made, and they are presented in Figures 4.1 and 4.2, the first was used for interviews with participants from Las Kellys Cataluña, and the second for interviews with participants from other associations of Las Kellys. The scripts were neither closed nor definitive but rather were used to identify key themes to be covered. Questions were written in an informal language register as it is important to align the language categories with those used by the interviewees. Interviews were transcribed in Spanish and then translated to English for this chapter.

Both the questions and the answers have been grouped to analyze and understand the path followed by Las Kellys, as well as how and why Las Kellys Cataluña decided to undertake as a solution to their setbacks. This approach attempted to detail the intended operation of the booking platform and the opinion of other

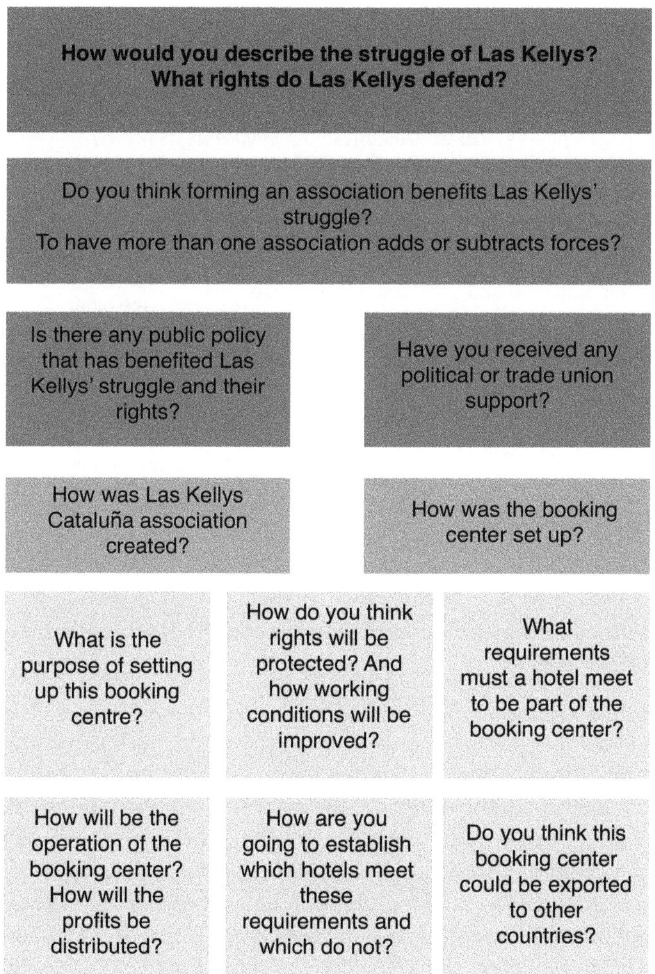

Figure 4.1 Diagram showing the script to interview *Las Kellys Cataluña*

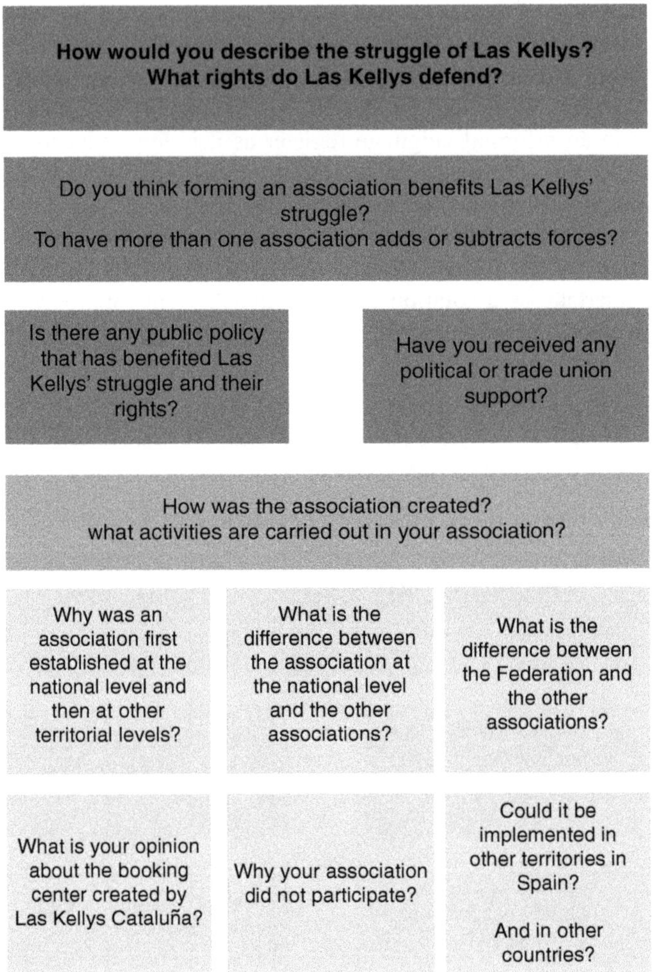

Figure 4.2 Diagram showing the script to interview the rest of associations that are not *Las Kellys Cataluña*

associations about that platform, as well as other possible forms of entrepreneurship that may be being conducted. In this way, five key themes have emerged: (1) the rights that Las Kellys are pursuing; (2) the activities and the funding of these association to pursuing such rights; (3) the relationship between Las Kellys itself and with political parties and trade unions; (4) the objectives that have led Las Kellys Cataluña to create the hotel booking platform and the plan to manage this platform; and (5) the opinions of Las Kellys that are not part of Las Kellys Cataluña about this booking platform, as well as the knowledge of other forms of entrepreneurship that could be being launched by any of Las Kellys to fight for the rights. The information was therefore organized and classified based on these five themes.

Interviews were conducted between December 2021 and August 2022; they lasted between 1 hour and 1 hour and 50 minutes and all were conducted by the first author. All were made by telephone and recorded with the permission of the interviewees. The associations were contacted through social networks, and through email. To our knowledge, from the 18 legally formed associations, there are two associations that are not functioning at the time the interviews were conducted, despite being legally inscribed. This includes the Madrid association, and the first created association at the national level. From the remaining 16, 5 associations agreed to be included: *Las Kellys Cataluña*; *Las Kellys Málaga*; *Las Kellys Granada*; *Las Kellys Sevilla*; and *Las Kellys Baleares*. The remaining 11 associations are functioning but did not reply to our emails.

Interviews were conducted with the chairwomen or vice-chairwomen of each of the five participating associations; the chairwomen or vice-chairwomen being the persons who often also act as spokespersons for the association. They are the ones who usually carry the weight of the entire organization and who know best how it works. It will not be specified which association has given each information or opinion, given that some information involves other associations, trade unions, or political parties, and authors consider that it may be too delicate to specify such information. The exception being specific information on the creation and development of the booking platform that, obviously, is information from Las Kellys Cataluña.

Results

The Struggle of Las Kellys, What Rights Do Las Kellys Defend?

Interviewees defined the struggle of *Las Kellys*: "A struggle to defend our labour rights"; "A desperation, desperation makes us take to the streets to protest, join in social networks, and create associations"; "A tireless, unjust, feminist, vindicative struggle that represents working women and women in all their facets… They have taken so much from us that they have taken away even our fear". And they affirm that "A very vulnerable group that needs work to survive is being deceived, this is labour exploitation, labour violence"; "Forming associations have helped women workers to know that they have their rights and that they have to claim them".

The main demands of Las Kellys are the following: "Ending subcontracting", "To put a workload limit, because it is the workload that makes us take pills, get sick, and have health problems that are not recognized"; "We are afraid of being fired…but that is a lie, we cannot be fired for not cleaning a specific number of rooms… The association is trying to let the hotel housekeepers know that this is illegal"; "We want that the plan of prevention of occupational risks is done with a gender perspective…"

They admit that it is not easy since "We are a group of vulnerable women… with personal burdens, mortgages" and "Everything is bigger than us, until creating the association was bigger than us, we do not know how things work, we do not have much idea… all the formalities are an odyssey…". In addition to working, the time they spend in the association "Is exhausting, the threats, the disputes…" But they believe that in the end "It has been a very positive struggle, since the movement began many things have been achieved", and that they are "Like a brotherhood, it is very beautiful, it requires sacrifices, but it has many rewards, together we have been advancing".

Activities Conducted by Associations

The activities the associations conduct could be summarized with this paragraph:

> We have focused more on helping our colleagues, informing them of our rights, knowing in which hotels the work situation is worse, and on visualizing our problems. In some cases, we also consult with the trade unions that allow us to work with them.

For these purposes they do "Meetings… informative talks…protests at the door of hotels"; "We train the hotel housekeepers…many do not know what is a trade union, a collective labour agreement, what are their rights"; "We reach agreements with the hotels on the workload, we look for work to those hotel housekeepers that have bad working conditions in their hotel"; "We made consults and solve doubts through social networks".

They also conduct other activities, for example, selling t-shirts, badges, and wallets; they have a great activity through social networks and even want to "Publish a book about the history of the association, the personal stories of some hotel housekeepers". They have even crossed borders, meeting with associations of France, Germany, or EEUU when the Marriott hotel chain announced that it would eliminate cleanliness of hotel rooms.

Funding of the Associations

None of the associations interviewed receive any public financial help. A few associations do have membership fees; however, these fees are voluntary. "We cannot charge a membership fee, it is not ethical, we are to help"; "We voluntarily put €3 or €5 if we have to make a trip"; "We have a symbolic fee of €30 per year, but if you cannot afford it, you don't pay it…Even if you are not part of the association and do not pay the fee, any housekeeper hotel that has a problem can come to us".

To finance themselves they sell t-shirts, the initial investment for the t-shirts was put in by themselves. "Travel and other expenses, such as lawyers and labour advisory, are paid out of our pocket". In addition, one of them has created a 'resistance box' in case of strike, to be able to make indefinite strikes, "as on those days of strike our salary is not paid, we want that 'resistance box' to help the colleagues who go on strike and cannot afford not to get paid".

One of the associations has raised money in a very peculiar way, thanks to "A theatre play called 'Kellys' for which we have given our image, and also appeared in a video clip for which they have given us money".

Some have had the luck of having "A lawyer who collaborates and advises us selflessly, if we have to go to trial the lawyer does charge us but gives us a discount".

> We have a labour advisory that does not charge us anything, even if we have to go to trial, but when we win a case, they charge us 15% of that compensation. They even provide us their training room for the meetings of the association.

Relations with Other Associations of Las Kellys

In general, participants shared that

> Having several associations adds forces, it is not the same to have an association in your own province as at the national level... Each province also has its labour agreement, so I could not counsel a woman from another province.

"Associations are safe places for housekeepers, I wish there were more associations, there is nothing more important than women themselves organizing, that is our strength". One participant stated that the reason for having different associations is that "We have always independence, we do not want anyone to tell us what to say or what to do". When the activity or demand is at the national level, they feel more comfortable doing it with the NGO Platform (Pecapis) since "The Platform is much more open, we work together, but each association is independent"; "The Platform allows us to work as we want, it is more flexible, we feel more comfortable...".

One of the participants did appear to be more critical, in admitting

> Among us there were problems, first we created the national association, but since there were those problems, then other associations were created. The Federation was created in 2019, but there were also problems, and it has not had a power of movement or decision...The fact that there are many associations and they do not work harms us all, this is why we created the Platform (Pecapis)... But the Platform is a very volatile because it has no rules, no organization, no legal structure, so it serves little purpose but to publicize our activities. ...Our problems have been more among women, who have taken this as a change of work, a way of feeling important, for their own benefit and not for the collective.

Institutional Relations

Most participants shared the sentiment that

> Politicians have supported us only to get votes, but the reality is that they have not had the will to solve the problem of subcontracts in the labour reform (refers to the amended of the article 42.1 of the 2015 Spanish Workers' Statute, already explained in the previous literature);

"Politicians are 'puppets' of the employers"; "They've used us as photos to win elections…, they still do not respect us as workers, neither as associations"; "We are like a grain in the ass and they make us promises, but they do not fix the problems".

Two participants were less critical and admitted that, although they do not do everything they promised, "Thanks to the meetings, the eventual hiring has been prohibited in Spain". Whilst this point regarding hiring was noted by two participants, it is important to note that eventual hiring has not, in fact, been prohibited. Rather, before the labour reform approved in 2021, the eventual hiring was a common practice of hiring in the Spanish labour market. Now it is only allowed in very specific cases due to justified circumstances of production or by substitution of worker (BOE.Es—BOE-A-2021–21788, 2021). Some participants did feel that the relationship between associations and politicians was positive:

> We have pretty good relationships overall. This year they have passed a law that the beds of hotels (a total of three hundred thousand beds) must be changed in the next 6 years to auto liftable and rotating beds with an 80% public subsidy to do so…

Participants were also quite critical of the trade unions: "We fight for our rights because there are no trade unions that protect us… Las Kellys born from the forgetfulness of the trade unions, politicians and employers"; "Some trade unions have played the role that they were there for us, and others have gone against us because they are afraid that we take away their merits, and they think we are hindering"; "Most of the hotel housekeepers do not want to know about trade unions because they do not understand us, in addition to the legal part, we need psychological support".

However, some participants also shared that "The role of the trade unions is necessary, since they are the only stakeholder recognized by law to negotiate with the hotels";

> We need the trade unions because the association cannot, for example, call a strike. I do not want to be critical of them because it is a collective that is afraid and there is nothing that weighs more than fear. But the trade unions are tired of us.

They also admit that "We support the trade unions if they support us". Whilst one participant went as far as to state that "Thanks to the trade unions many improvements were made in the conditions of the hotel housekeepers in the last hotel collective agreement… They are the ones who are going to negotiate our future, we cannot throw stones at them".

The Kellys Cataluña claim that they became a trade union in 2018 because "In 2017 the collective labour agreement of Catalonia was negotiated, and we wanted to negotiate that agreement but as an association we could not negotiate". Some of the associations admit that "We have thought about being a trade union, it is a debate that we have", but others shared that "We have never thought about creating a trade union because we believe that this work is already covered, there are already trade unions, we must not divide more".

Hotel Booking Platform Centre Entrepreneurship

It was the ambivalence felt towards politicians and the trade unions that generated the impetus for the hotel booking platform:

> Seeing that politicians never did anything… and after approving the Seal of Quality and Fair work for the tourism and hospitality sector by the Parliament of Catalonia, we found another wall of difficulties… The Seal of Quality was not put into practice, the employers began to put up a fight because to obtain the Seal no worker can be outsourced, and for us, without the fundamental point of prohibiting outsourcing, that Seal was useless…It was late 2021 and the Seal was still not implemented, we began to look for other solutions and we remember that people always ask us for a list of fair hotels, hotels that respect the rights of workers…Then we thought about getting the Seal of Quality and from there comes the idea of setting up the hotel booking centre…We spent more than 6 months planning everything and it came out thanks to all the people who supported us and who are aware that we must create a new world…It is a paradigm shift, we want to create a clean world, outside of corrupt politicians and governments that only serve to rob the people and not to fulfil their promises to solve people problems.

On their website the Catalonia association states their objectives of the platform, which are quite broad and ambitious (Goteo, 2021): This reservation platform

> Aims to be a tool for social transformation that goes beyond booking accommodation in hotels. We want to clean up the world and inaugurate the new era of tourism based on respect, beauty, and well-being, which puts human interests before mercantile interests around the planet. We will demonstrate that putting human interests before mercantile interests is profitable, generates wealth, happiness, peace, trust and love in people.
>
> (Goteo, 2021)

> The reservation platform will promote responsible consumption, only hotels that meet the requirements of the Seal of Quality and Fair work could be reserved. This will promote and create decent work, which is why we will succeed in ending poverty, hunger, and social and gender inequality. Workers will also enjoy health because one of the requirements of the Seal of Quality is the respect to the law of prevention of occupational risks…We will also impose ecological cleaning products that respect the health of workers and the environment.

The requirements to belong to this hotel booking platform are the same as those requested to obtain the Seal of Quality. The association admits that "not many hotels meet these requirements, but there are some". The association plans to verify through audits how hotels can meet the requirements, although they have not yet fully established how it will be done. The answers to the questions regarding the operation of the platform were vague. In respect to whom will be responsible for audits, they claim that "We already have some people, it will be someone who has already worked in hotels, they have to be independent, fair, with a little humanity, does not have to belong to Las Kellys Cataluña, but yes support us". The association will develop an app to make reservations, "the app is still being developed by computer professionals". They do not know how many people they have hired, they are "few, they are not many, that is why we go slow". They believe that with the amount they have obtained through crowdfunding will be enough to pay the start-up costs, given they have obtained €90,235.

Several aspects remain unclear. For example, it is not known how hotel commissions will work, "Well… We get commissions like Booking.com…I do not know which is usually charged". And regarding the benefits, it was indicated that "They will be used to pay the staff, and the workers of the hotels will also earn a part". But it is not clear why the workers of the hotels will earn part of the profits of the hotel booking platform. Other participants explained that

> The money obtained thanks to the crowdfunding will be destined to pay IT costs of the reservation website, the mobile application, the legal advice, technical advice on processes and financing the secretariat and translation. As well as advertising, organizing events, travel, and daily subsistence allowance in other countries to establish alliances and take the initiative to other countries. In addition, they plan to organize annual events in which awards are given to the cleanest and most respectful hotels on the planet.
>
> (Benito, 2021)

There were times when *Las Kellys Cataluña* were concerned that they would not reach the minimum requested (€60,000)

> We have passed a horrible stress, but in the end, it was worth it, it was a very nice experience…There are many committed people who already want to use the platform and who write to us to know when it will come out.

The association's objective is to implement this platform not only in Catalonia:

> Since we started with this project, some hotels have written to us, even from outside Catalonia, but for now we will start with Catalonia. The idea, however, is that it expands throughout the planet and that everyone can use this booking centre and book rooms through Las Kellys.

Opinion of Other Associations of Las Kellys about the Entrepreneurship

The other associations admit not knowing very much about the implementation of this booking platform. With the opinions of participants from other associations being quite different: "Since they are strong, Barcelona is a big city, they are also a trade union". Another affirms: "Everything that any association does, that they believe that they must do, and that they decide to do seems perfect to me...", but "A company offered us to create something similar, but we said no, because they just wanted to benefit from our struggle. I do not know what it was like for them. If they do it independently, I applaud them". And there were participants who shared doubts: "I would not know what to tell you, on the one hand, I am okay with that, but on the other hand, we are not a travel agency...but if it goes well, bravo!". One participant was extremely critical:

> I see it as absurd...for me it has been practically a scam, they have made a lot of money and I think it will be for nothing... they did it through a hotel company, so they are divested... I do not see that the creation of the trade union is right either...Everything that Las Kellys Cataluña are doing I see it quite strange, I do not understand how they can get so much money.

All the associations believe that the launch of the hotel booking centre has a fundamental problem: it is not possible to launch a reservation centre with two or three hotels: "They have all the hoteliers against them, who will work with them?" and, on the other hand, ignorance: "They have no idea what they're doing".

In developing this enterprise to other territorial levels, participants from other associations claimed that "It could be done, I guess, but it is not our function, we do not have a commercial purpose and we do not have the resources to do so". Whilst others questioned the potential to expand geographically:

> Here few hotels meet the requirements, it would be only two or three hotels that we could include. And nationally there are lists of hotels made, but every day you are taking hotels off the list because they stop fulfilling some requirement.

One participant believed that a roll out in all territories is not necessarily required

> I do not see it for all places, here the agreements are fulfilled, we are not outsourced, hoteliers are paid days off. We do not have to go to that extreme of creating a booking centre to put pressure on hotels.

Rather than aligning with the potential of the online booking platform, other associations were creating their own initiatives: "We have an ethical list of 'Kellys friendly hotels' that we publish every year and try to reach the citizenry". Another participant noted

> We are working with a team on a Seal of Quality in employment and we are also creating a booking centre working with hoteliers. But it has been a lot of work and a lot of negotiation with the hotels ... In addition, considering all employees of the hotel, not only Las Kellys...Also, hotels must meet the 2030 Agenda... In addition, the hotels would not be charged commissions

> per sale, they would be charged for the installation and maintenance of the program, audits and so on, but not the commissions, and those commissions that they could save would serve to improve the quality of the work…Our booking centre is almost finished, but we need more hotels that want to participate because with two or three we cannot launch a booking platform.

And another participant indicated that "What is being studied now is to create Seal of Quality for hotels". Important, however, whilst there was a lot of optimism attached to the potential of these various initiatives, the pandemic has paralyzed this work because the setback in working conditions has forced the associations to refocus their efforts on protests, demands, and colleagues' support.

Discussion

The Las Kellys movement is a movement of women who have decided to come together with their own means and knowledge to fight for their rights and to try to overcome the shortcomings and gaps in the economic, political, and social system as well as the inactivity at times of political parties and trade unions. As presented in the literature review, women's associations are created not only to support and help other women with their problems but also to transform the political, social, and economic context of society (Freund & Hernandez-Maskivker, 2021). Las Kellys are probably the best example of this, and they are associations that have been created not only to help each of their members individually but also to change the work paradigm of an entire collective.

Las Kellys meet all the previously analysed motivations that lead to entrepreneurship such as demanding and inflexible work environments, work-life balance, lack of recognition, or occupational segregation (Carvalho et al., 2018; Costa et al., 2016; Noguera et al., 2015; Ribes-Giner et al., 2018; Russen et al., 2021). And yet, even though they are women who knew nothing about laws, associations, and workers' organizations and who, without financial funds, have been able to overcome all the barriers and constraints to be an entrepreneur presented in the literature review such as lack of investment or access to advisors, thus corroborating the assertions that network membership can help women entrepreneurs overcome constraints and barriers (Kimbu et al., 2019), partly due to the selfless help of lawyers and labour advisories and because the social and media support they have obtained (Alcalde-González et al., 2021) that in many cases has helped them not only as moral support but as advice, collaborating altruistically with these associations or even contributing money. In the face of difficulties, the associations of female hotel housekeepers in Spain, instead of retreating, have been learning to fight for their rights and have won great victories. Las Kellys have overcome the limitations that have been presented in previous literature which suggest that their networks are limited (Ribeiro et al., 2021) and now the society, the political parties, trade unions, and employers know who Las Kellys are and know that they are willing to sacrifice their time and money to continue fighting for the rights of the collective of women who make up the hotel housekeepers in Spain.

In fact, some have decided to go a step further, through enacting entrepreneurship by creating their own ethical hotel booking centre. The main objective of this type of entrepreneurship is, in the first place, to force hotels to meet the same requirements that ought to be met to obtain the Quality Stamp approved by the Catalonian Parliament in 2018. A Quality Stamp that was never implemented but politically recognized as central to the sustainability of the sector. This would improve the working conditions of hotel housekeepers and, in turn, allow them to have more time to demand that governments implement social policies such as access to early retirement or recognition of their ailments as occupational diseases. And, secondly, to change the employment paradigm of an entire sector, extending it to other countries, in line with the literature review that states that women's associations strive for a transformation of society (Ribeiro et al., 2021).

The way to get funding for the start-up of the platform has not been easy, and so far, it cannot be stated that there is a clear roadmap, plan for implementation, or knowledge of how to generate income and employ employees; nor is their knowledge of how profits are going to be distributed. The development and maintenance of a platform requires a high investment, plus the salary of the workers. It seems difficult they can obtain a level of sales high enough to bear the expenses, especially considering that the commissions per sale of hotels do not usually exceed 15%. In addition, with bed banks suppliers buying hundreds of rooms from hotels, it is unlikely that hotels will worry about not appearing in the booking platform. It is thus a project that lacks development and has drawbacks already pointed out by some of the Las Kellys associations not involved in the platform's development.

It is also recognized that few hotels meet all the requirements, and not every hotel will want to collaborate with Las Kellys. And this inconvenience has difficult solutions. The only possible solution comes from the society, without the commitment of a society that seeks another type of consumption, who feel responsible for a change in society and who are concerned about the working conditions of the hotels in which they stay, they will hardly be able to operate any initiatives. When female entrepreneurs connect with the society, they can deal better with the economic and regulatory environment (Neumeyer et al., 2019). If the society is involved in that paradigm shift that Las Kellys claim and commit to reserve through this booking centre, even having few hotels in its catalogue, it is possible that more and more hotels try to meet the requirements.

With all these inconveniences it is early to assert if it is possible to export this entrepreneurship to other territories And even more if we consider that the relations between the associations themselves are not always easy and that in some cases they prefer to work independently. It is presumed difficult to create 18 platforms, and the only viable solution would be to create a national booking centre that had in its catalogue hotels from all over Spain and managed this booking centre by a higher-level group, as the Federation or the NGO Platform, on behalf of all associations. It may be better, as noted by some participants, to create a Seal of Quality in employment at the national level among all associations and to grant the hotels that meet the requirements with that Seal, as one more quality certificate. Create a list of the hotels that obtain this Seal and distribute that list among society, tour operators, and travel agencies. Those tour operators and travel agencies that commit themselves to book these hotels to their customers could also receive the seal. As the hotel distribution system is organized, that would be the best way to point out those hotels that meet the requirements of *Las Kellys*, counting on all the members of the hotel distribution chain.

Conclusion

Even if it does not ultimately achieve the desired success, this entrepreneurship will be a further step in the struggle of the female hotel housekeepers in Spain and a further demonstration of the ability of the female collective initiatives to transform both their members and society (Freund & Hernandez-Maskivker, 2021). Projects like cooperatives, social entrepreneurships, non-profit entrepreneurship, or "mompreneurship" have reformulated the hegemonic androcentric definition of entrepreneurship (Serrano-Pascual & Carretero-García, 2022). As Cañada (2015, p. 42) states in his book: Strengthening the empowerment of workers in the tourism sector is fundamental if we aspire to another tourism. Create a responsible consumer movement that pays attention to the working conditions of the hotels where they stay. Establish alliances between trade union organizations, works councils, and citizens' movements. Las Kellys as group and as association could be empowered; in fact they have demonstrated from the beginning a huge fortress to put their demands on the political and social agenda.

At the personal level, hotel housekeepers are women who want to enjoy their family, their children, and their hobbies but who have not always been able to do so fully due to a job that squeezes their physical and mental health (Otegui Carles et al., 2023, p. 21). When women cannot enjoy their life, their family, or their hobbies because of a job, they can hardly feel empowered. However, at a group level, empowerment has indeed occurred; Las Kellys have demonstrated from the beginning a huge fortress to put their demands on the political and social agenda. As a result, they have achieved, as has been presented throughout the chapter, changes in labour legislation, with the inclusion of some of their illnesses in the catalogue of occupational illnesses or with the modification of temporary contracts in the labour reform; in working conditions, whether due to the understanding with employers or to the legal battles won; in the way they are treated by politicians and trade unions, being received by both groups to talk and debate about work and rights; and even in the way they are seen by society and the media, where they have become recognized stakeholders in the hospitality sector.

Both the fulfilment of some of their demands and the non-fulfilment of others have contributed to their growing empowerment as a group, taking another step forward with the entrepreneurship presented in this chapter and with the rest of initiatives such as quality seals or lists of ethical hotels still to be developed. If successful, the "hotel-housekeepreneurship" presented in this chapter could be a model of entrepreneurship that could be transposed to other countries, changing society through a change in the form of tourism consumption that considers the working conditions of workers. At the same time, it could remove barriers in the advancement of gender equality in tourism through a sustainable entrepreneurship that could lead towards a more sustainable future (Butkouskaya et al., 2020).

The biggest limitation is that both the entrepreneurship of Las Kellys Cataluña and the other initiatives that are being implemented by other associations of Las Kellys are still under development. Therefore, as future lines of research, it is proposed to start from this chapter to investigate how these entrepreneurial performances unfold in ways that have consequences in the fight for the rights of hotel housekeepers in Spain.

References

Alcalde-González, V., Gálvez Mozo, A., & Valenzuela Bustos, A. (2021). No clean rooms, no hotel business: Subversion tactics in Las Kellys' struggle for dignity in hotel housekeeping. *Annals of Tourism Research, 91*, 103315. https://doi.org/10.1016/j.annals.2021.103315

Benito, C. (2021, July 29). Las Kellys abriran una Central de Reservas Hoteleras con sello de trabajo justo y de calidad. *Catalunya Plural.* https://catalunyaplural.cat/es/las-kellys-abriran-una-central-de-reservas-hoteleras-con-sello-de-trabajo-justo-y-de-calidad/

BOE.es—BOE-A-2021-21788. (2021). *Real Decreto-ley 32/2021, de 28 de diciembre, de medidas urgentes para la reforma laboral, la garantía de la estabilidad en el empleo y la transformación del mercado de trabajo* (testimony of Jefatura Estado). https://www.boe.es/buscar/act.php?id=BOE-A-2021-21788

Butkouskaya, V., Romagosa, F., & Noguera, M. (2020). Obstacles to sustainable entrepreneurship amongst tourism students: A gender comparison. *Sustainability, 12*(5), 1812. https://doi.org/10.3390/su12051812

Cañada, E. (2015). *Las que limpian los hoteles. Historias ocultas de precariedad laboral.* Barcelona: Icaria Editorial.

Carvalho, I., Costa, C., Lykke, N., & Torres, A. (2018). Agency, structures, and women managers' views of their careers in tourism. *Women's Studies International Forum, 71*, 1–11. https://doi.org/10.1016/j.wsif.2018.08.010

Catalunya, P. de. (2018). *Moció subsegüent a la interpel·lació al Govern sobre les condicions laborals dels treballadors del sector turístic.* Parlament de Catalunya. https://www.parlament.cat/web/activitat-parlamentaria/siap/index.html?STRUTSANCHOR1=detallExpedient.do&criteri=302-00026/12&ad=1

Costa, C., Breda, Z., Pinho, I., Bakas, F., & Durão, M. (2016). Performing a thematic analysis: An exploratory study about managers' perceptions on gender equality. *The Qualitative Report.* https://doi.org/10.46743/2160-3715/2016.2609

Delgado, G. (2018). *Las kellys, mujeres obreras se organizan contra la explotación y el maltrato laboral.* https://www.mundoobrero.es/pl.php?id=7732

Figueroa-Domecq, C., de Jong, A., Kimbu, A. N., & Williams, A. M. (2022). Financing tourism entrepreneurship: A gender perspective on the reproduction of inequalities. *Journal of Sustainable Tourism, 32*(3), 656–676. https://doi.org/10.1080/09669582.2022.2130338

Figueroa-Domecq, C., de Jong, A., & Williams, A. M. (2020). Gender, tourism & entrepreneurship: A critical review. *Annals of Tourism Research, 84*, 102980. https://doi.org/10.1016/j.annals.2020.102980

Freund, D., & Hernandez-Maskivker, G. (2021). Women managers in tourism: Associations for building a sustainable world. *Tourism Management Perspectives*, *38*, 100820. https://doi.org/10.1016/j.tmp.2021.100820

Goteo, F. (2021). *Yo reservo con Las Kellys*. Goteo.org. https://www.goteo.org/project/las-kellys

Hsieh, Y.-C., Apostolopoulos, Y., & Sönmez, S. (2013). The world at work: Hotel cleaners. *Occupational and Environmental Medicine*, *70*(5), 360–364. https://doi.org/10.1136/oemed-2012-100986

Hsieh, Y.-C., Apostolopoulos, Y., & Sönmez, S. (2016). Work conditions and health and well-being of Latina hotel housekeepers. *Journal of Immigrant and Minority Health*, *18*(3), 568–581. https://doi.org/10.1007/s10903-015-0224-y

Kellys. (2015, December 11). Quiénes somos. *LAS KELLYS*. https://laskellys.wordpress.com/quienes-somos/

Kimbu, A. N., Ngoasong, M. Z., Adeola, O., & Afenyo-Agbe, E. (2019). Collaborative networks for sustainable human capital management in women's tourism entrepreneurship: The role of tourism policy. *Tourism Planning & Development*, *16*(2), 161–178. https://doi.org/10.1080/21568316.2018.1556329

Kourtesopoulou, A., & Chatzigianni, E. (Elia). (2021). Gender equality and women's entrepreneurial leadership in tourism: A systematic review. In M. Valeri & V. Katsoni (Eds.), *Gender and tourism* (pp. 11–36). Emerald Publishing Limited. https://doi.org/10.1108/978-1-80117-322-320211002

Ministerio del Interior. (2023). *Consulta del Fichero de Denominaciones de Asociaciones |: Sede Electrónica del Ministerio del Interior.* https://sede.mir.gob.es/opencms/export/sites/default/es/procedimientos-y-servicios/tramites-sobre-asociaciones-inscritas/consulta-del-fichero-de-denominaciones-de-asociaciones/

Mooney, S., & Ryan, I. (2009). A woman's place in hotel management: Upstairs or downstairs? *Gender in Management: An International Journal*, *24*(3), 195–210. https://doi.org/10.1108/17542410910950877

Moral-Martin, D. (2020). ¿Por qué se han organizado las camareras de piso? Algunas claves e interpretaciones desde la revitalización sindical. *Revista Española de Sociología*, *29*(3-Sup1), Article 3-Sup1. https://doi.org/10.22325/fes/res.2020.59

Navarro, F. J. C. (2020a, April 29). *Las 'kellys', más fuertes que nunca: Nace la Plataforma Estatal de Camareras de Pisos*. Otro Periodismo. https://otroperiodismo.com/las-kellys-mas-fuertes-que-nunca-nace-la-plataforma-estatal-de-camareras-de-pisos

Navarro, F. J. C. (2020b June 5). *La Plataforma Estatal de Camareras de Pisos solicita amparo al Defensor del Pueblo*. Otro Periodismo. https://otroperiodismo.com/la-plataforma-estatal-de-camareras-de-pisos-solicita-amparo-al-defensor-del-pueblo

Neumeyer, X., Santos, S. C., Caetano, A., & Kalbfleisch, P. (2019). Entrepreneurship ecosystems and women entrepreneurs: A social capital and network approach. *Small Business Economics*, *53*(2), 475–489. https://doi.org/10.1007/s11187-018-9996-5

Nimri, R., Kensbock, S., Bailey, J., Jennings, G., & Patiar, A. (2020). Realizing dignity in housekeeping work: Evidence of five-star hotels. *Journal of Human Resources in Hospitality & Tourism*, *19*(3), 368–387. https://doi.org/10.1080/15332845.2020.1737770

Noguera, M., Alvarez, C., Merigó, J. M., & Urbano, D. (2015). Determinants of female entrepreneurship in Spain: An institutional approach. *Computational and Mathematical Organization Theory*, *21*(4), 341–355. https://doi.org/10.1007/s10588-015-9186-9

Otegui Carles, A., Fraiz Brea, J. A., & Araújo Vila, N. (2023). Living and working as a female hotel housekeeper before, during, and after the Covid-19 pandemic. *Journal of Human Resources in Hospitality & Tourism*, *22*(3), 435–459. https://doi.org/10.1080/15332845.2023.2180956

Ribeiro, M. A., Adam, I., Kimbu, A. N., Afenyo-Agbe, E., Adeola, O., Figueroa-Domecq, C., & Jong, A. de. (2021). Women entrepreneurship orientation, networks, and firm performance in the tourism industry in resource-scarce contexts. *Tourism Management, 86*, 104343. https://doi.org/10.1016/j.tourman.2021.104343

Ribes-Giner, G., Moya-Clemente, I., Cervelló-Royo, R., & Perello-Marin, M. R. (2018). Domestic economic and social conditions empowering female entrepreneurship. *Journal of Business Research, 89*, 182–189. https://doi.org/10.1016/j.jbusres.2017.12.005

Rosemberg, M.-A. S. (2020). Health and safety considerations for hotel cleaners during Covid-19. *Occupational Medicine, 70*(5), 382–383. https://doi.org/10.1093/occmed/kqaa053

Russen, M., Dawson, M., & Madera, J. M. (2021). Gender diversity in hospitality and tourism top management teams: A systematic review of the last 10 years. *International Journal of Hospitality Management, 95*, 102942. https://doi.org/10.1016/j.ijhm.2021.102942

Serrano-Pascual, A., & Carretero-García, C. (2022). Women's entrepreneurial subjectivity under scrutiny: Expert knowledge on gender and entrepreneurship. *Gender, Work & Organization, 29*(2), 666–686. https://doi.org/10.1111/gwao.12806

5 Women Entrepreneurs Sense-Making of Policy Supports in Context of Rural Tourism in Kazakhstan

Aliya Tankibayeva, Sandy MacDonald and Akbota Saduakassova

Introduction

Tourism in Kazakhstan is a growing sector, where rural tourism is envisioned to maintain the viability of vast rural territories and maintain the well-being of local communities (Grandi et al., 2022). Measures to foster rural tourism include expanded public policy and private sector involvement. Although in a nascent stage of active experimentation, rural tourism has already revealed its potential. The Lonely Planet (Coffey, 2022) guide included Kazakhstan in Best in Travel 2021 Community Accommodation featuring the development of a locally managed, sustainable form of tourism homestay infrastructure in Kazakhstan. A year earlier, in 2020, the British Backpacker Society listed Kazakhstan among the best five must-visit destinations for adventure travel (Brady, 2020).

In the tourism sector, women entrepreneurs in Kazakhstan mainly operate micro and small enterprises and rarely enterprises of medium and large size (Mynbayeva, 2016). The extant literature about women entrepreneurs is scarce and fragmented. However, it provides evidence that women's entrepreneurial choices and strategies are subject to gender-related influences (Abou-Shouk et al., 2021; Filimonau et al., 2022). Women in rural areas are particularly vulnerable as entrepreneurs due to challenges associated with social approval and overall low experience in entrepreneurship, lack of start-up capital, and restricted access to credit (Kulekeyev et al., 2019). Whereas policy supports are available, it is unclear how effectively the needs of women entrepreneurs are addressed and what mechanism enables women to identify and participate in policy measures. Further, the COVID-19-induced disruption to tourism has mobilized various support interventions, the effectiveness of which is yet to be understood. This chapter aims to contribute to understanding effective policy design and delivery by exploring women's decision-making while seeking and enacting available policy support. That is, the main research objective is to develop a systemized understanding of women entrepreneurs' transformative choices in response to the available policy support. Although this chapter analyses policies initiated before the pandemic, exploring women's sense-making will also be helpful in a more nuanced understanding of the micro-foundations of policy in rural tourism for post-pandemic recovery.

DOI: 10.4324/9781003315858-5

This chapter is structured into four sections. The *Literature Review* contains an overview of the current situation of rural women entrepreneurs in the tourism sector in Kazakhstan and the tourism-related policy context. It concludes with an explanation of the Theory of Planned Behaviour (TPB; Ajzen, 2020) as applied to explore women's sense-making. *The Research Method* section explains the choice of data collection and sampling approach, and approach to qualitative data analysis. Further, the *Results* section elaborates on women's intention to either participate or disregard policy support derived from their sense-making. Last, *the Discussion and Conclusions* section assesses the relevance of TPB and enunciates implications for the current policy enhancement.

Literature Review

Women's Entrepreneurs in Tourism Sector

In line with the trend worldwide, women-led businesses are usually concentrated in specific sectors of the economy, requiring less starting capital and tending to expand slower (Koellinger et al., 2013). In 2022, over 1.5 million SME enterprises were operating in Kazakhstan, and over 700,000 of them were run by women (BNS, 2022). There are traditionally 'female' sectors of the economy like wholesale and retail, accommodation and food services, education, transactions with real estate, healthcare, and other services, where women comprise over half of the SMEs (Satpayeva, 2019). Also, the leading share of the informal sector, or something which is known as the self-employment sector in Kazakhstan's economy, lies in the abovementioned enterprises (Satpayeva, 2019).

While women in Kazakhstan have actual opportunities to realize their rights in the majority spheres of social life (Table 5.1), women in rural areas might be found less advantageous. Xheneti et al. (2019) emphasized that local contexts play a determining role in women's motives and success in becoming entrepreneurs. In the survey undertaken by the Applied Economics Research Centre in 2017, on an average rural inhabitants spent about 49.5% of their income on food, with the overall income in rural areas spent for consumption about 25% less than in cities (Kulekeyev et al., 2019, p. 21). At the same time, for 20% of women-respondents – their work is the primary source of income for their family, and 10% of families live off their wife's allowance or pension (Kovyazina, 2020). Women spent about three times more time on unpaid homework than men –3.5 hours vs. 1.1 hours, remarkably less time on networking, leisure, and sports activities –2 as compared to 3 hours, and slightly less time on socializing –41 minutes vs. 49 minutes and personal hygiene –12.2 hours vs. 12.5 hours (BNS, 2023). Notably, unemployed women (28%), mothers with many children (46%), and housewives (47%) expressed low confidence in the potential of rural women to develop a successful enterprise, while working women outside entrepreneurship were most confident (Kulekeyev et al., 2019, p. 47). Indeed, Filimonau et al. (2022) identified the share of women entrepreneurs in Kazakhstan who are in a prolonged vulnerable life situation – the context characteristic to women in rural and remote areas.

Table 5.1 Women's Economic Empowerment and Equality (WE3) 2021, USAID

Indicator (0–5, Higher Is Better)	*Kazakhstan*	*World*	*Data Used on:*
Financial services score	2.63	2.3	Access to credit, bank account, credit cards, and debit cards
Ownership score	2.63	2.62	Property rights
Technology score	2.55	2.61	Internet and mobile phone usage
Business score	3.19	2.53	Time, cost, and procedures related to starting a business
Decision-making score	1.76	2.51	Household responsibilities and laws on divorce
Percentage of individuals using the internet, female	90%	–	Females who used the internet from any location in the last three months
Global gender gap: Index score (0 to 1, higher is better)	0.72	–	Gender-based gaps on dimensions: Economic Participation and Opportunity, Educational Attainment, Health and Survival, and Political Empowerment

The Policy Context for Women's Entrepreneurship

Policy support for women entrepreneurship in Kazakhstan is diverse, drawing on objectives and instruments from national and local policies for the economy, specifically entrepreneurship, social and employment policies, and diverse programmes on women's leadership. Namely, 'Auyl - El-Besigi', 'Development of agro-industrial complex', aims for economic modernization of rural areas, while programmes 'Business Road Map-2030' and 'Program of mass entrepreneurship development' highlight support opportunities for SMEs. In the tourism sector, the *Concept for the Development of the Tourism Sector* highlights entrepreneurship among key drivers and objectives of sector development. Other supports are available through the social assistance policy or ad-hoc programmes such as COVID-19 recovery support. Women are entitled to social payments, for instance, due to pregnancy and childbirth, childcare, disability, and loss of breadwinner. Although coverage is universal, the communication on procedures and eligibility varies in the extent of ease of understanding and targeted delivery (PM, 2023).

For better reachability of the policy measures, there are 207 state employment centres, one-stop informational windows operated by local administration for residents in regional cities, more remote settlements (Enbek, 2023), and non-stop online service. Nonetheless, in the survey, only 50% of the unemployed rural women knew about policy support measures, and among the self-employed – about 60% (Kovyazina, 2020). Both levels of awareness and participation are low. Only 4%–9% participated in support programmes, 42%–48% heard something but did not participate, and about 60% never heard about opportunities for micro-loans, entrepreneurship or business skills training, employment support, or free professional education.

Mainly, state policies and measures are gender-inclusive, i.e., endorsing equal access regardless of gender. At the same time, specific supports designed for women have been set up by sectoral interest groups, NGOs, businesses, and collaborations of governments with other stakeholders. National policies, in the majority, provide financial support and delegate capacity building to non-state operators. Some conditional grants are also provided by other development entities. Nonetheless, women report difficulties meeting the eligibility requirements for financial support due to the absence of collateral or small-scale entrepreneurial activity. FIn 'Bastau Business' the women's acceptance rate for micro-finance loans in 2021 was about 31% (Myrzakhmetov, 2021). In general, where the support programmes involve large amounts of capital or target investment into fixed assets – women tend to be two to three times less active than men (DAMU, 2020). On the one hand, this is consistent with the literature evidence that women prioritize responsible investment behaviour (Baig et al., 2021). However, there is also evidence of structural and cultural constraints women experience when attempting to level up from micro- and small entrepreneurship (Yacus et al., 2019; Cowling et al., 2020; Brindley, 2005).

A plausible conclusion is that women in rural areas encounter a gender-related challenges – as women (taking up child- and family-care roles), as entrepreneurs (seeking social approval, need for balancing family and business responsibilities, and increasing the viability and scale of enterprise - from micro-enterprises to small and medium) and as rural residents (access to information and resources). In this complex policy terrain, insights into women's sense-making illuminate how and why women apprehend policy support as worth taking.

Theory of Planned Behaviour

According to the theory, when the intention is more substantial, an individual is more prone to undertake a behaviour to fulfil that intention. Evidence in the literature supports the importance of the intentional component for acting as entrepreneurs (Sarwar et al., 2021; Zampetakis et al., 2017; Kautonen et al., 2013); however, it is scarce on women's intentions to participate in policy support (Asare et al., 2022; Cooke & French, 2008).

The Theory of Planned Behaviour considers an individual's perceived behavioural control, attitudes, and subjective norm as the most critical predictors of the intention

(Ajzen, 2006; 2020). *Attitudes* are beliefs shaping either favourable or unfavourable evaluation regarding the behaviour of interest – in this chapter – available policy supports. Attitudes are contingent on *subjective norms* – individualized perceptions of social approval or disapproval directed at the behaviour of interest. The current study is interested in exploring whether there is a contradiction in declared societal norms calling to participate in policy programmes and women's experience in their life situations.

Further, TPB allows to deliver insights regarding why and how women substantiate their movement from intention to behaviour. It conceptualizes *perceived behavioural controls* as a self-efficacy factor that manifests the strength and character of women's capacity to identify and utilize the resources and capability to act in a constrained situation (Baluku et al., 2020; Yoopetch, 2020). To various extent, attitudes and subjective norms incorporate social influences and thus manifest structural and socio-cultural influences internalized by women. Combined, attitudes, subjective norms, and perceived controls define women's *intention* to participate in the policy support, while perceived behavioural controls can moderate the effects of subjective norms in intention formation through adjusting a woman's self-concept regarding feasible behaviour that she is capable of (La Barbera & Ajzen, 2020).

Research Method

The research setting is Almaty region, Turkestan region, West Kazakhstan region, North Kazakhstan, and East Kazakhstan region. While Turkestan and West Kazakhstan have generally more traditional roles assigned to women rather than Almaty, North and East – women in rural areas of all five regions carry similar family and child care, housekeeping, and similar extent of decision-making and economic powers (Kulekeyev et al., 2019). A purposive sample of 40 women entrepreneurs was recruited for the study via snowball sampling and from a pool of researchers' contacts (Table 5.2). The data were collected in 2022 and the first half of 2023 using semi-structured and unstructured interviews. Due to the relatively small sample, the results are not fully applicable to an advanced level of theorizing. Nevertheless, they are sufficient to capture patterns of intention formation by rural and self-empowerment by rural women concerning specific types of policy support.

To trace the effects of the type of policies on women's intentions and choices – each interviewee was asked only about one type of programme, although some participated in more than one type of support. Five women respondents did not take part in any support programme. The interview data was analysed through thematic analysis (Braun & Clarke, 2006) where general TPB themes were sought and identified through the iterative analysis process of most persistent concepts, that is, initial codes and noting the patterns of their interaction (Table 5.3).

Table 5.2 Sample Profile

Policy Support Taken	*Respondent*	*Age*	*Children/ Other Dependants*	*Years in Entrepreneurship (If Applies)*	*Entrepreneurial / Professional Activity*	*Type of Enterprise*	*Region*
Entrepreneurial training (national framework)	A	38	Married, two children, husband disabled	8	Herbal products for wellness, homestays	SP	EK
	B	52	Extended Family	10	Apiary, honey products, tasting, and informational tours	SP	AR
	C	29	Single mother, one child	5	Sport and festive event organization	SP	TR
	D	56	Single	About 20	Homestay	SE	TR
Women-targeted training entrepreneurial training	E	28	Single, parent family	4.5	Pastry and souvenir sweet making	SP	AR
	F	48	Married, non-monetary support from husband, disabled child	5	Vegetable mini farm, agri-production	SP	AR
Tourism product/ service development training	G	34	Single, non-monetary support from family	7	Souvenir production, master classes	SP	NK
	H	36	Husband passed away 2 years ago	6	Guest house	SP	TR
	I	38	Married	About 15	Riding School	FE	NK

(*Continued*)

Table 5.2 (Continued)

Policy Support Taken	*Respondent*	*Age*	*Children/ Other Dependants*	*Years in Entrepreneurship (If Applies)*	*Entrepreneurial / Professional Activity*	*Type of Enterprise*	*Region*
Product/Service / Enterprise development Grants	J	41	Extended family, children, grandchildren	Less than 1	Nature-based tourism – tours and sightseeing	SP	WK
	K	31	Married, one child, non-monetary support from husband	6	Eco-auyl (eco-village)	FE	AR
	M	39	Married, two children	2	Dairy Production	SP	WK
Concessional loans or loan guarantees	N	36	Two children	5	Wellness services	SP	EK
	O	40	Married, one child	6 years in current enterprise, 10 years' service	Kid care centre	SP	AR
	P	38	Married, two children	4 years	Poultry mini farm	SP	NK
Women-targeted enterprise/product or service grants	Q	29	Parent family, husband's parent family, non-monetary support from husband	5 years	Mini eco-hotel, honey, and herbal products	SP	EK
	R	49	Widow	Less than 1 year	Dairy production	SP	NK
	S	44	Married, three kids	14 years	Event organizations	SP, LLP	WK
	T	30	Non-monetary and monetary support from partner	6 years	Bike-sharing, homestays for cyclists	SP	NK

Employment assistance programme	U	36	Married, three children	n/a	Training in pattern- and dressmaking	unemployed	AR
	V	26	Single	n/a	Registration for job placement	SE	AR
Social payments for COVID-19 disruption	W	39	Single, two children	7 years	Catering	SP	TR
	X	25	Single, parent family	3 years	Nurse, baby-sitter, cleaning, etc	SE	AR
	Y	44	Widow, three children	n/a	Nurse, baby-sitter, cleaning, etc	SE	WK
Business/ Entrepreneurial mentorship programmes	Z	40	Married, four children, initially a breadwinner, non-monetary support from husband	12	Greenhouses	Farm	CK
	AA	38	Married, two children, non-monetary support from husband	11	Mini eco-hotel, wellness products, wellness tours	SP, LLP	EK
	AB	27	Single	6	Traditional crafts (leather)	SP	TR
	AC	32	Married, non-monetary support from husband	3	Traditional cloth and dressmaking	SP, FE	AR

(Continued)

Table 5.2 (Continued)

Policy Support Taken	*Respondent*	*Age*	*Children/ Other Dependants*	*Years in Entrepreneurship (If Applies)*	*Entrepreneurial / Professional Activity*	*Type of Enterprise*	*Region*
Start-up incubation programme	AD	52	Married, children, grandchildren	22	Traditional dressmaking	SP	WK
	AE	38	Married, one child	9	Traditional jewellery-making	SP	NK
Trade and supply chain facilitation/ partnership development programmes	AF	58	Married, children, grandchildren	23 in current business, 35 overall	Agricultural production	SP, LLP	TR
	AG	35	Married, non-monetary support from husband	4, 11 years overall in entrepreneurship	Eco-farm	SP	AR
	AH	47	Married, non-monetary support from husband and son	7	Production of national dwellings and furniture	LLP, FE	WK
Decided not to participate	AI	24	No	1	Catering at tourism sites	SP	TR
	AM	40	Married, main breadwinner	8	Mini eco-hotels, camping	SP	WR
Decided not to participate, did not qualify	AJ	45	Two children	15	Tour guide, nature-based tourism	SE	AR
Did not qualify	AK	23	Single	3.5	Educational tours for schools	SP	AR
	AL	36	Single, two children	10	Restaurant and bakery		TR

SP, sole proprietorship; SE, self-employed; FE, family enterprise; LLP, limited liability partnership.

Table 5.3 Coding Frames

Main	*Themes and Codes*	
Needs and motives	• necessity • aspirations • rural development • national development • economic empowerment • curiosity and novelty • infrastructure and connectivity (constrained)	
Intention formation	Attitudes	Policy • quality of information/communication (awareness) • trustworthiness • reliability • relevance to local context • relevance to tourism sector Self • anxiety • capability • peers (support, information, results) • self-restriction for not too high expectations
	Subjective norms	Social • declared norms • actual encounters • gendered effects • structural effects Perceived • socially approval/disapproval • conventional roles Empowerment strategies • relying on rural identity • multiple role balance • incremental change • conventional support circles • new support circles • role-models (distant and proximal)
	Perceived controls	Material • land and natural resources • financial support Non-material • family support • community support • peer support • hard work • time • personal choices and autonomy • confidence • learning and knowledge • communication and interpersonal skill • service (tourism product) • travel/mobility

(*Continued*)

Table 5.3 (Continued)

Main	*Themes and Codes*	
Behaviour	Policy participation • feasible in local context • capable of fulfilling • tailored to personal circumstance	
	Empowerment	Economic • scaling-up • closing investment gaps • new product development • income generation Psychological • self-efficacy • supporting others • control over own learning and skill acquisition Socio-cultural • balancing multiple roles • transformational impact for gender roles • positive impact to others • positive impact to local area

Results

Needs and Motives

First, women perceive empowerment motives as present in all combinations of needs they expressed regarding the existing policy. Even when 'necessity'-driven, the motives were accompanied by hope, readiness for new experiences, and a strong aspiration for success.

> It was a time of testing. My son was seriously injured, and I had to leave my job, and financial support from relatives was not enough. I did not want to sit, get upset and wait for help from anyone, I understood that I can only count on myself, and I wanted my son to be happy and I could respect myself for what I achieved. Giving up was inacceptable for my(F).

They mention both the need and desire to attain a 'better life' economically and to rise in social status. The second frequent motive was to realize one's own 'dream' of establishing an enterprise while contributing to a community and nation, and the third was to fulfil one's curiosity and search for new experiences.

Results further suggest that women recognize their areas' remote location and rural character – the rural-centric identity motivates women to invest in their area and be proud of their activity's positive impact. Most respondents want to stay in their places and are particularly motivated to contribute to better conditions for their area alongside fulfilling their own goals.

> Over the years of activity, I received letters of thanks from the Akim of the city, from the Council of Businesswomen for my contribution to the sustainable development of the rural area. I do this because I care about what is happening in my country. I want to help my people and my ayul (village).
>
> (M)

> In fact, Kazakh land is very rich, gives us so many blessings, give us everything if we take good care of it. I can't imagine abandoning it. My husband and I believe we do not have moral right to degrade the land or add harmful chemicals to the products - this land should be clean for us and for our children.
>
> (AF)

Women expressed that they expect policy measures to address their local needs: low and intermittent employment, help for child care and medical care, and ensuring regular energy and water supply. Seventeen respondents (approx. 41%) perceived their ambition and behaviour as constrained by such context (Respondents U, AJ,) – however, accumulating a certain threshold of courage, they can initiate entrepreneurial activity and apply for policy support. While national economic and tourism sector development policies are trusted, the willingness to participate is moderate for the perceived difficulty with application and capability of fulfilling the obligations. Programmes of smaller scale and with less formal requirements, for instance, independent initiatives or tourism-specific training and small grants, are perceived with positive curiosity and hopefulness.

The main discouraging factor for those who decided not to participate in policy support was the doubts regarding whether the policy was well-targeted for women's circumstances and capabilities:

> I realized I could not apply alone. In addition to my plant, I must build storage, while also to earn more than now to repay the loan for both plant and storage. Should we gather five-ten women and work on the storage - my proposal would be feasible.
>
> (AG)

Respondents considered whether it is possible to realize a proposed support instrument given the actual cost of renting or buying tourism premises, cost and availability of utilities, availability of roads, transport, and communication, and how their participation will be commensurate with the family commitments.

Intention Formation

Attitudes

Personal presuppositions based on prior life history and awareness of policy designs affected women's decisions whether to participate in the policy support.

For their choice-making, women consider experience and advice from whom they know by relating efforts, risks, and chances for success as commensurate with their local context. In contrast, distant role models are sought for inspiration. Notably, women tend to self-restrict their hopes and lower expectations from the policy. Partly, they do not believe in their success, while partly they feel uncomfortable and anxious to change their life circumstances radically. Respondents to emphasize the strong effect from the policy personnel on their attitudes from the informants and policy personnel with whom they directly interact.

> I was doubtful at first. Competition for the whole country, with all those who are at the beginning of the path of thousands, if not hundreds of thousands. Is it realistic to receive one of a dozen grants? But I looked at how previous winners work and what they say and decided to take a chance. It was not easy to change the attitude of doubt, but once the goal is set - there I had energy, and my advice to all 'Do not lower your own goals - take steps!
>
> (K)

Women further suggest that their engagement, particularly in tourism, allows them to encounter visitors of various backgrounds and places – thus enhancing their mindsets, knowledge about the sector and life choices, and acquiring social skills:

> After studying and communicating with peer-women in our training class, I decided to live to the maximum, that is, to give myself to love, goals, and children, to realize myself in my activities, to fill my life with meaning, and become ready for change. I also understood that there are many opportunities to live happier and resources around that I could not see before. We, rural women, are used to being modest and not expecting much - but we have many things to share.
>
> (AC)

In this way, tourism entrepreneurship for women is associated with empowerment not only because of economic opportunities and skill development but also because of helping women to become happier and more fortunate, and open-minded in life.

Subjective Norms

Women respondents perceive a disconnect between the declared social norms and women's practical encounters with social approval and disapproval while becoming or acting as tourism entrepreneurs. The remarkable controversy is attributed to strong community approval for prioritizing family roles and following traditional family hierarchies. The respondents suggested that their entrepreneurial and autonomous activities are socially better accepted when these activities are carried out for women or in domains conventionally associated with women. Interestingly, women mention many assertive and autonomous behaviours in those conventional roles: for example, being 'active', 'enthusiastic', 'influential', 'a good organizer', 'negotiator', 'problem-solver', and 'brave and energetic'. However, women do not feel comfortable negotiating with men and, therefore, tend to focus on tourism products and goods where it is possible to combine them with conventional roles of mothers, wives, housekeepers, family members, etc.

> Women are good at many tasks for running a business, however, I prefer to work with other women - it's difficult to negotiate with men, or to start a product where men usually take the lead. When I am more established as a business, I think I should be more confident and try.
>
> (P)

Women-oriented programmes are perceived better when they are framed as economic programmes or rural development for women rather than leadership or rights-oriented programmes. Nonetheless, women respondents expect some objectives or potential outcomes regarding gender equity, equality, and inclusivity from all policy support. At the same time, women seek policy measures that consider their desire to combine multiple gender-specific and other roles in ways suitable for their rural life:

> In women's hands is the destiny of the nation. In our family, the role of women is to glue the family together, to help children grow healthy and able. But if a woman is not realized, not able to herself - what can she pass on to the children? I recommend all women to seek opportunities to learn and apply for support - it is ok to start with small steps, and then the family will only appreciate your success and support you.
>
> (AD)

All respondents acknowledge the effect on subjective norms from role models, either distant or whom women know personally. The most frequent type of role model described by women in the sample tends to represent an 'active', 'respected', 'hardworking', 'honest', 'courageous', 'candid', 'courteous', and 'happy' woman who combines work and family and who has remained a 'rural' woman whose empowerment is realized with acknowledgement of rural context. Respondents indicate that fears of negative consequences and change anxiety can be self-imposed due to a lack of experience in choice- and decision-making:

> Myself before the project and after are two different people - I was insecure, and after training, I became a woman who learned to solve my problems and extract new opportunities from any crisis. Because of participation, I revealed my organizational strengths and other abilities which I was self-limiting or doubting before.
>
> (E)

Subsequently, communication and interpersonal skills are mentioned for raising confidence to change unfavourable communal perceptions of their entrepreneurial activity. Participation in the programmes include both formal and informal learning components to develop those skills and real networks of supporters.

Perceived Controls

Respondents unanimously identify land as an essential source of confidence – they believe that fertile lands are an asset providing them with nutrition and other means of living. It is perceived as a reliable anchor that is not subject to fashion, given hard work and proper treatment. Their ability to work on the land and to process agricultural goods and their skills in crafts enhance women's confidence against difficult life situations.

> In tourism, I am a novice, but I know how to treat the land and work hard. On your land, you are free to do what you want - it's a great blessing but also a responsibility. I am grateful that I could have learned about the innovation to cultivate it and acquire a loan for them.
>
> (J)

In this light, women respondents (79%) tend to trust policies that have understandable links with available local resources or clear ways to acquire new resources. Not willing to relocate, most respondents however would like to apply for policy support with more opportunities to visit other places in the country and abroad.

Second, women mention support from the family as an immaterial resource increasing their capability for entrepreneurship. In families, besides moral support, women who are not otherwise eligible for bank services or policy requirements for financial support can acquire initial funding from extended family and friends. Nonetheless, while financial resources and results were positively associated with confidence to act, respondents suggested they were anxious about possessing and managing finances autonomously:

> Initially, I acted intuitively, very vaguely having an idea about what exactly I needed to do. It was very stressful because I had never managed a business or undertaken a large investment. I realized that my choices were only mine and I had to make all the decision, so I undertook many rounds of questions and answers and learned from mentors, neighbours, other winners, organizers, asked for support from the centre and the Chamber.
>
> (O)

They suggest that observing other women managing choices and undertaking incremental changes in behaviour helped them develop a mindset of entrepreneurs with their own resources and finances. Thus, policies offering mentorship and direct encounters with role models were mentioned to be more attractive.

Respondents who did not participate in policy support indicated that they either needed help to orient themselves in the application procedures or perceived a lack of control over critical subjective and objective resources:

> I understand that one day I will be able to reach the eligibility, but as for now, I decided not to risk it. I looked carefully and realistically, my guest-house is functioning so far, and just for reaching eligibility, I do not want to speed up and stretch over what I have now. I work all day and get tired and cannot understand how to work more and what exactly will be a good decision to scale up. First, I need good ideas about what will be good for my enterprise.
>
> (AM)

For all respondents, issues related to their health and low access to social infrastructure were perceived as difficult to control. Such aspects as childbirth or treatment of acute conditions often take time away from their settlement and increase women's dependence on support from the family or significantly diminish energy and available time for other activities.

> I start blaming myself that I am not sufficiently trying - but she [program coordinator] said that taking care of children while they are newborn is full-time work; and that women have different circumstances and different supports - with me, it was not possible at that moment, but I could try later.
>
> (AJ)

At the same time, all respondents indicate that social assistance programmes are available and well-communicated, and the number of social payments provided noticeable help to combine daily routines and entrepreneurship. However, difficulty in controlling their own time and desired activities makes them feel dependent, anxious, and constrained.

Behaviour: Empowerment and Participation in Policy Support

Women expect transformational effects from their participation in policy support. However, respondents mentioned that, in their experience, not all women-specific programmes integrate rural-oriented features into their designs. Programmes oriented toward rural development followed by mentorship and networking programmes were mentioned as most accommodative to help develop tourism products and enhance business skills. Overall, respondents were satisfied with the impact of participation in support policy, especially if they could qualify and receive enterprise development grants. The absence of commitment to repay the grants made women feel more secure and focused on the aspects of product and enterprise development in which they considered themselves most capable – hence, more impactful:

> For a small, even a mini-micro-enterprise like mine, it would be difficult to raise income within the time required by large loans, so I must choose either go for a micro-loan or take a contest for the grant. I do not regret the time filling out the application and the contest results took about seven months. Now

> that I do not have to worry about repayment - I could focus on my product and researching customer needs.
>
> (M)

Respondents appreciate that financial support was closing critical needs for investments in capital assets, construction of facilities, and acquisition of technology and equipment – without which their enterprises were likely not to obtain resources necessary for starting up. Financial commitments stimulate women to learn structured and systemic goal setting, financial planning, sales, legal and technical knowledge, and launching better marketable products. Notably, skill acquisition and financial support are most valued by respondents when women perceive them as enabling them to be simultaneously more authentic and entrepreneurial and making a positive impact on the community:

> I want to lead a more authentic lifestyle, to live in harmony with nature and my inner self. I already have a guest house; I plan to expand the settlement so that I can share this healthy and happy experience with my guests. Many want to return and recommend their experience to their friends and so on. I see the big difference on their first day here and after some time: they can rest, eat healthy, and maybe their life is better after that.
>
> (Q)

In line with the subjective norms – among the desired positive outcomes that women expect from their participation in support policy is to learn to balance multiple roles: business, professional, family, and non-work, and non-family commitments:

> To combine family and business will require efforts from both the woman and the man. In life, for the most part, it happens as follows - women begin to compromise on their goals and desires, believing that only one thing can be achieved, and they begin to suffer from the fact that part of their aspirations and desires are not realized. When I started my enterprise, I was sure that I wanted to be a mother and wife, a tourism professional, and an entrepreneur.
>
> (AA)

Knowledge exchange and peer support during and after formal participation were mentioned to have empowering effects not only on women's perceived controls but on entrepreneurial and leadership behaviours. For several respondents, encouraging other women entrepreneurs became essential to their motivation to grow as an enterprise:

> For example, neighbours come, and we start mini-consultations - how much time does she need, how many preserves or where to find childcare, and when best to go to the city…I am proud that my house became a centre for women entrepreneurs' support, and some come with their requests while some come

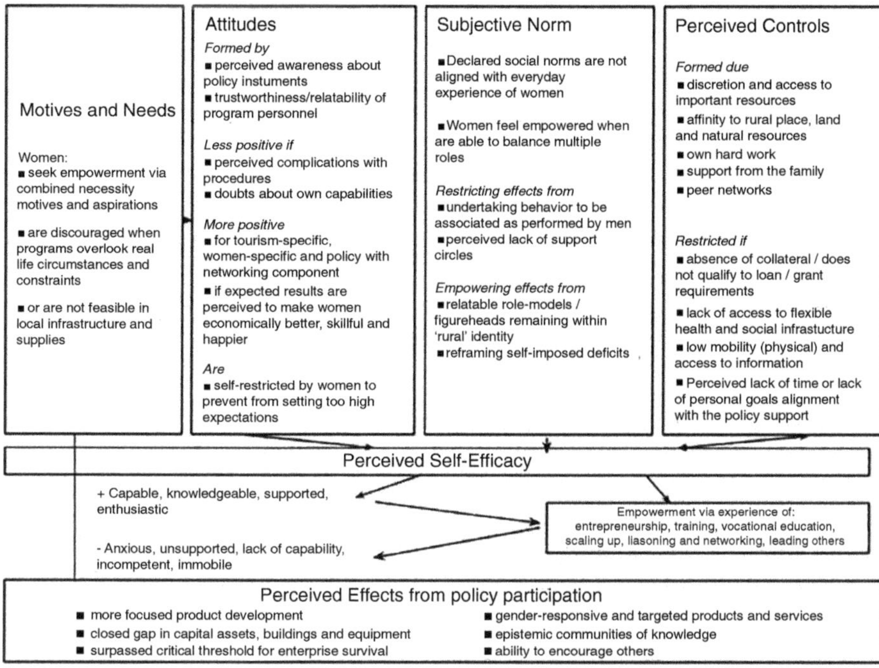

Figure 5.1 Women empowerment through policy sense-making and participation (compiled by authors)

> to share advice. When I see that this changes lives of many people for the better - I understand that I do everything right.
>
> (B)

Consequently, women respondents considered their ability to deliver results responsive to the needs of women in tourism – customers and other entrepreneurs to be an essential outcome of their participation in policy support and their entrepreneurship in general (Figure 5.1).

Discussions

Theory of Planned Behaviour

The results support the relevance of the Theory of Planned Behaviour in examining women's choice-making rationales and sense-making. For the exploratory study undertaken in this chapter, the theory captured a nuanced way in which women understand the sources of the constraining and enabling subjective norms they have regarding participation in policy support. TPB appeared to be well-accommodative for identifying, among others, the self-exposed components in subjective norms and perceived controls of resources, thus valuable for studies of subjective empowerment processes and effectiveness of learning acquisition from capacity-building.

The theory is also helpful in differentiating intention from behaviour and positioning perceived controls as an individual inventory for realizing the specific forms of behaviour. In addition to identifying the resources that individuals perceive critical for the feasibility of behaviour, it was possible to capture the importance of attitudes and values – simply possessing necessary resources and controls and the absence of constraining norms have not directly resulted in initiating life-changing choices. Women perform sense-making of the policy support by sensitizing their motives and possible consequences of the behaviour – inclusion of the latter can be suggested as an extension for theorists of TPB to consider.

Implications for Women's Empowerment in Rural Tourism

The results supported that self-efficacy is seen as an enabler, a goal, and an outcome of empowerment. At the same time, women evaluate not only the resources and support (material and socio-psychological) they have but also how the available resources are commensurate with their skills and knowledge to manage them. Even when resources are provided, women may perceive themselves as anxious or inexperienced using them. Hence, training components aimed at simultaneously mindsets and skills are crucial for women's participation in policies.

Women in the sample see their rural identities as fundamental to their empowerment and do not intend to abandon conventional support they perceive in their social context. While they are willing to undertake risks and innovations regarding their enterprises and learn new entrepreneurial competencies, they evaluate policy measures on contextual appropriateness. The rural tourism sector is perceived as particularly accommodating for their preferred empowering choices because rural identities can be integrated with developing entrepreneurial products and processes. Tourism-related activities involving learning from visitors have the potential for socio-psychological empowerment, which should be explored for policy design and implementation.

Implications to Policy Design and Communication

Currently, the relevant support to women entrepreneurs in the tourism sector of Kazakhstan is mainly received via rural development policies, while tourism-specific policies are regarded as frequently changing and fragmented. Awareness of the programme and how clearly the programme is communicated is crucial for women's decision-making. Moreover, the frontline personnel appeared significant in shaping women's awareness and attitudes to the programme.

Programmes and policy measures with educational components tailored for specific needs of rural women are positively evaluated. Further, the policies and programmes with relatable role models are perceived as more motivating for women to participate. Women entrepreneurs and to-be entrepreneurs in rural tourism seek policy designs that expand their social networks – this may take the forms of cross-visits or virtual platforms of peers or mentors for mutual support and knowledge exchange.

Women in the sample positively acknowledge that existing support policies have a multicomponent and cross-modular design allowing them to take a tailored route by selecting components or switching across components concerning their goals and time. Women emphasize that they seek the feasibility of policy instruments in their local context and their life circumstances. However, women positively evaluate the effect of the policy on their psychological and personal empowerment regardless of a chosen policy instrument.

Conclusions

This chapter applies the Theory of Planned Behaviour to explore the intention formation of rural women entrepreneurs to participate in a range of tourism support policies and empowerment rationales of women in relation to policy support. It was found that women expect empowerment in a form suitable to their rural context: while their intention is complexly formed, the choices for participation are mainly guided by commensurate perceived resources with their own capabilities to manage them. Implications to women empowerment and policy design and communication are discussed.

References

Abou-Shouk, M. A., Mannaa, M. T., & Elbaz, A. M. (2021). Women's empowerment and tourism development: A cross-country study. *Tourism Management Perspectives*, *37*, 100782.

Ajzen, I. (2006). *Constructing a theory of planned behaviour questionnaire (2011)*. Retrieved from https://people.umass.edu/aizen/pubs/tpb-faq.pdf (accessed March 16, 2020).

Ajzen, I. (2020). The theory of planned behaviour: Frequently asked questions. *Human Behaviour and Emerging Technologies*, *2*(4), 314–324.

Asare, M., Lanning, B. A., Montealegre, J. R., Akowuah, E., Adunlin, G., & Rose, T. (2022). Determinants of low-income women's participation in self-collected samples for cervical cancer detection: Application of the theory of planned behaviour. *Community Health Equity Research & Policy*, 0272684X221090060.

Baig, U., Hussain, B. M., Davidaviciene, V., & Meidute-Kavaliauskiene, I. (2021). Exploring investment behaviour of women entrepreneur: Some future directions. *International Journal of Financial Studies*, *9*(2), 20.

Baluku, M. M., Matagi, L., & Otto, K. (2020). Exploring the link between mentoring and intangible outcomes of entrepreneurship: The mediating role of self-efficacy and moderating effects of gender. *Frontiers in Psychology*, *11*, 1556.

BNS. (2022). *Number of registered small and medium-sized businesses as of 1 January 2022*. Agency for strategic planning and reforms of the Republic of Kazakhstan Bureau of National statistics. https://gender.stat.gov.kz/page/frontend/detail?id=106&slug=1-2018&cat_id=6&lang=en

BNS. (2023). *Average number of hours, spent on paid and unpaid housework (total labor burden), by sex*. Agency for strategic planning and reforms of the Republic of Kazakhstan Bureau of National statistics. https://gender.stat.gov.kz/en

Brady, S. (2020, January 6). The best country for adventure backpacking in the 2020s, according to experts. *Lonely Planet*. Retrieved May 14, 2024, from https://www.lonelyplanet.com/news/algeria-backpacking-adventure-destination

Braun, V., & Clarke, V. (2006). Using thematic analysis in psychology. *Qualitative Research in Psychology*, 3(2), 77–101.

Brindley, C. (2005). Barriers to women achieving their entrepreneurial potential: Women and risk. *International Journal of Entrepreneurial Behaviour & Research*, *11*(2), 144–161.

Coffey, H. (2020, November 17). Lonely Planet best places to travel 2021: the full list, from Kazakhstand to Gothenburg. *Lonely Planet*. Retrieved from https://www.independent.co.uk/travel/news-and-advice/lonely-planet-best-travel-2021-destinations-sustainable-kazakhstan-gothenburg-b1722641.html (accessed August 1, 2024)

Cooke, R., & French, D. P. (2008). How well do the theory of reasoned action and theory of planned behaviour predict intentions and attendance at screening programmes? A meta-analysis. *Psychology and Health*, *23*(7), 745–765.

Cowling, M., Marlow, S., & Liu, W. (2020). Gender and bank lending after the global financial crisis: Are women entrepreneurs safer bets? *Small Business Economics*, *55*, 853–880.

DAMU. (2020, January 1). *Indicators of gender statistics of entrepreneurship* [*Pokazateli gendernoj statistiki predprinimatel'stva*]. Retrieved 25 May, 2024 from https://damu.kz/poleznaya-informatsiya/informatsiya-fonda/analitika/

Enbek. (2023). *Employment centers* [*Centry zanjatosti naselenija*]. Retrieved May 26, 2023, from https://www.enbek.kz/ru/employment-services/czn

Filimonau, V., Matyakubov, U., Matniyozov, M., Shaken, A., & Mika, M. (2022). Women entrepreneurs in tourism in a time of a life event crisis. *Journal of Sustainable Tourism*, *32*(3), 457–479.

Grandi, S., Macdonald, S., & Tankibayeva, A. (2022). The role of knowledge structures in reconfiguring rural tourism in response to the Covid-19 pandemic: An exploratory study of rural tourism in Italy and Kazakhstan. *Geo Journal of Tourism and Geosites*, *41*(2), 555–563.

Kautonen, T., Van Gelderen, M., & Tornikoski, E. T. (2013). Predicting entrepreneurial behaviour: A test of the theory of planned behaviour. *Applied Economics*, *45*(6), 697–707.

Koellinger, P., Minniti, M., & Schade, C. (2013). Gender differences in entrepreneurial propensity. *Oxford Bulletin of Economics and Statistics*, *75*(2), 213–234.

Kovyazina, K. (2020). *Rural women are doubly vulnerable* [*Selskie jenzhiny uyazvimy vdvoine*]. Retrieved 24 May 2024 from https://ekonomist.kz/kovyazina/uyazvimy-vdvoine/

Kulekeyev, Z., Turekhanova, B., Tuleuov, O., Urazgaliyeva, A., Kovyazina, K., Orazbekova, S., & Bauyrzhan, A. (2019). Research of economic opportunities of rural women to consider their inclusion as a distinct socially vulnerable category into the Governmental economic agenda / Isslledovanie jekonomicheskih vozmozhnostej sel"skih zhenshhin dlja v jekonomicheskuju povestku pravitel"stva vkachestve otdel"noj kategorii social"no ujazvimogo naselenija, Applied Economics Research Centre. Soros Kazakhstan Foundation (rep.). Retrieved from https://aerc.org.kz/storage/app/media/projects/final_rus.pdf (accessed August 1, 2024)

La Barbera, F., & Ajzen, I. (2020). Control interactions in the theory of planned behaviour: Rethinking the role of subjective norm. *Europe's Journal of Psychology*, *16*(3), 401.

Lonely Planet. (2020, October 20). *Lonely planet's best in travel 2021*. Lonely Planet Publications.

Mynbayeva, J. (2016). *Women in Kazakhstan*. London: Wonder Foundation Insight Report.

Myrzakhmetov, A. (2021). *Less than third of applications for Bastau Business loans approved*. Retreived 25 May, 2024, from https://kaztag.info/en/news/less-than-third-of-applications-for-bastau-business-loans-approved-

PM. (2023). *Employment, financing of business initiatives and social benefits: what support women receive in Kazakhstan*. Retrieved 25 May, 2026 from https://primeminister.kz/en/news/employment-financing-of-business-initiatives-and-social-benefits-what-support-women-receive-in-kazakhstan-23339

Sarwar, A., Ahsan, Q., & Rafiq, N. (2021). Female entrepreneurial intentions in Pakistan: A theory of planned behaviour perspective. *Frontiers in Psychology*, *12*, 553963.

Satpayeva, Z. (2019). *Zhenskoe predprinimatel'stvo v Kazahstane: sostojanie i perspektivy* [*Women's entrepreneurship in Kazakhstan: Status and prospects,* Center of Gender Economics Research, Narxoz Univerdity -Soros –Kazakhstan*, Almaty online from* https://www.soros.kz/ru/%D0%B7%D0%B0%D0%B9%D1%80%D0%B0-%D1%81%D0%B0%D1%82%D0%BF%D0%B0%D0%B5%D0%B2%D0%B0/

USAID. (2023). *Idea*. Retrieved May 26, 2023, from https://idea.usaid.gov/cd/kazakhstan/information-and-communications-technology-ict

Xheneti, M., Madden, A., & Thapa Karki, S. (2019). Value of formalization for women entrepreneurs in developing contexts: A review and research agenda. *International Journal of Management Reviews*, *21*(1), 3–23.

Yacus, A. M., Esposito, S. E., & Yang, Y. (2019). The influence of funding approaches, growth expectations, and industry gender distribution on high-growth women entrepreneurs. *Journal of Small Business Management*, *57*(1), 59–80.

Yoopetch, C. (2020). Women empowerment, attitude toward risk-taking and entrepreneurial intention in the hospitality industry. *International Journal of Culture, Tourism and Hospitality Research*, *15*(1), 59–76.

Zampetakis, L. A., Bakatsaki, M., Litos, C., Kafetsios, K. G., & Moustakis, V. (2017). Gender-based differential item functioning in the application of the theory of planned behaviour for the study of entrepreneurial intentions. *Frontiers in Psychology*, *8*, 451.

6 Social Policies for Supporting Women Entrepreneurs in the Tourism Industry in Kenya

Rehema Mugendi-Kiarie

Introduction

Tourism has been touted as a major foreign exchange earner for developing countries and its position in the economies of the developing countries has been elevated. As an example, earnings from tourism in Kenya in the year 2021 amounted to Kenya Shillings One Hundred and Forty-Six Billion (Kshs. 146.5 billion),[1] which was higher by 65% in comparison to the year 2020 (Statista, 2022). Tourism is viewed as one of the industries that is experiencing rapid growth with a high potential to contribute to the overall gross domestic product (GDP). In South Africa, the government has labelled tourism as a key driver of economic development and creation of jobs with a contribution of approximately 11% to South Africa's GDP (Deen, Harilal, & Achu, 2016). Tourism contributes to an estimated 6%–7% of both direct and indirect jobs worldwide, while one in twelve workers in the world are engaged in the travel and tourism industry (Ward, 2010).

Women make up half of the population in the world while at the same time contribute to the production of 37% of the world's GDP (Statista, 2015). The deliberate exclusion of women in economic activities results in increased poverty levels since half the population in an economy is excluded from gainful activities resulting in inefficiencies in the use of resources (USAID & Scott, 2021). Furthermore, the unbalanced power and unequal involvement of women and men in decision-making are some of the key issues that lead to social and political instability that ultimately results in poverty (Swiss Agency for Development and Corporation [SDC], 2003). To build a strong foundation for a successful economy and just society, it is necessary to empower women for them to take part in economic development.

The focus of this chapter is on women in the Maasai Community, a pastoralist community occupying a vast semi-arid area in the County of Kajiado. Within the Maasai Community, women are involved in micro and small enterprises that produce ornaments and garments for the local and foreign tourists. With changes in climate, cattle herding, the main activity of the community has been affected by lengthy durations of drought resulting in the depletion of water resources and grazelands that have led to livestock deaths and loss of sources of income (Tugjamba, Walkerden, & Miller, 2023). In addition, cattle rustling, conflict arising from scarce

DOI: 10.4324/9781003315858-6

water resources and pasture has made it impossible for women to take care of their families (Young, 2012). Apart from the government and development partners assisting these communities, the need for a sustainable source of income is vital for the future sustenance of their livelihoods. Empowering the local communities, more so women, through entrepreneurship reduces their dependence on government interventions while at the same time providing a source of livelihood. By encouraging the local communities to participate in community groups to engage in gainful economic activities that benefit all, the key essence of social entrepreneurship is developed and passed on from generation to generation within the community.

This chapter seeks to answer two fundamental objectives, namely:

1 To uncover the factors that influence the empowerment of women entrepreneurs in the tourism industry in Kenya.
2 To examine how social policies can enable women entrepreneurs to improve their livelihoods and those of the tourism communities through entrepreneurship.

To address the above objectives, this chapter draws on the STEEPLE framework, an adaptation of the PESTEL framework, which has been shown to be effective for understanding the external factors that affect tourism activities (Seyitoglu & Ivanov, 2020; Mayaka & Prasad, 2012). This framework is relevant because it enables women entrepreneurs influenced by the external factors that are likely to hinder business activities and, for which action is needed, to cushion themselves against any adverse effects that may occur. STEEPLE factors form a great learning point for any entrepreneur since whatever happens in the larger macro-environment will affect the businesses at the micro-environment level.

This chapter also looks at cultural tourism, the cottage industries operated by women in the Maasai Community and the empowerment that is needed to enable women entrepreneurs to contribute to social tourism. The women are tourism entrepreneurs because they take risks and invest their resources to participate in income-generating activities in order to fill market needs around the tourism sector or destination (Kimbu & Ngoasong, 2016) With cultural tourism, this study has explored some of the cultural traits that draw tourists to these communities and the economic benefits that these communities have been able to reap as a result of using culture to enhance their entrepreneurial spirit.

The chapter is structured into the following sections: a literature review has been undertaken focusing on the tourism sector and the STEEPLE model; the methodology for undertaking the study has been described including the methods of data collection and the target population; the results from the data collection exercise have been tabulated and categorized into themes; a summary of the findings has been discussed including the linkages between the findings and the STEEPLE model and the chapter ends with a discussion and conclusion on the social policies for supporting women in the tourism industry in Kenya.

Literature Review

Empowering Women to Reduce Poverty and Increase Economic Growth

Research has shown that in the developing countries, an estimated 60% of women are in non-agricultural informal employment (ILO, 2002). Proponents have advocated for gender equality in tourism through supporting women by providing decent employment, empowering women economically through focusing on business tourism, educating and training women on achieving success, providing women with an avenue to take leadership and make decisions and involving the community and civil society to support women and address inequalities in gender at home and in the communities (UN Women, 2020).

A World Bank (2007) study illustrates how empowerment women can contribute to reducing poverty and increasing economic growth. The study suggests that increasing gender equality in the community and business results in more women participating in the labour force and being the caregivers at home, their children fare much better. In addition, there is also a reduction in the poverty levels and growth in the economy. Increasing women's earnings results into greater investment in the education, health and nutrition of their children that results in long-term economic growth (Gendernet, 2011).

In Africa, there are more women than men who choose to become entrepreneurs. However, women entrepreneurs face more challenges, specifically: inequality, lack of capital and discrimination, which results in earnings that are lower by an average of 38% in comparison to enterprises owned by men (Gaye, 2018). Despite the challenges faced by women entrepreneurs in Africa, success stories have been recorded in various countries. In Senegal, the resilience of women in the tourism industry has been illustrated through a small rural camp, Chez Léontine, that is run by a woman entrepreneur who dropped out of school and had to endure stereotyping in the community whereby those close to her only considered her fit enough to run a family. Despite the lack of resources, in addition to other challenges, her small tourism firm has steadily grown (Santander Fundación, 2020). In Cameroon, women entrepreneurs are actively involved in the tourism industry exploring locally available opportunities and resources to run their own small tourism firms at the local level (Kimbu & Ngoasong, 2016). The immense contribution made by women in the economy either through entrepreneurship, agriculture, employment, or unpaid homecare needs to be recognized. Subsequently, gender equality, inclusive economic growth and poverty eradication can occur through investing in women's economic empowerment (UN Women, n.d.).

The STEEPLE Model and the Tourism Industry

The STEEPLE (social, technological, economic, environmental, political, legal and ethical) analysis tool is an adaptation of the PESTLE model that is utilized to assess, evaluate and review the macro business environment to assist in making decisions that influence business development (Saidova & Muminjonova, 2022).

The tourism industry is volatile, and any changes in the external environment have an amplifying effect on the performance of the tourism industry. Entrepreneurs, government, and civil society need to understand and identify the risks that are posed by the various factors that operate in the macro-environment to respond to any adverse effects. The following external factors associated with the STEEPLE model have been explained in relation to the tourism industry.

Socio-cultural factors: These factor takes two forms; tourists are likely to travel to a particular country to experience its unique culture. On the other hand, travel and tourism has in recent times been viewed as fashionable. As a result, the popularity of travelling has led to more individuals opting to travel as opposed to staying at home (Bush, 2019).

Technological factors: Travel technology has been widely embraced, and it includes the use of IT and e-commerce to facilitate travel, tourism, and hospitality activities. The use of technology in travel and tourism reduces amount of time spent on sourcing for destinations, cost reduction in selection of accommodation and transport. It also serves as an important tool for marketing and establishing affordable travel destinations within the budget of the tourists (Revfine, 2021).

Economic factors: Economic factors are related to taxation, inflation, exchange rates and the available disposable income. Countries that have lower tax rates for tourist-related activities are more attractive since tourists can spend less for more. The world over, as the levels of disposable income increase, more people are willing to travel and spend on tourism-related activities. Favourable exchange rates for tourists offer more value for money.

Environmental factors: Despite the increase in travel and tourism activity, transport pollution is one of the negative outcomes due to the use of modes of travel that are still reliant on fossil fuels. Waste generated by tourists as they travel including littering and contamination is also another negative side effect. Another major form of waste is water. A lot of water is utilized for tourist activities, e.g. hotels, golf courses, swimming personal and personal water uses. Furthermore, failure to put in place water recycling plants in most hotel facilities has been detrimental to the environment.

Political factors: The political instability of a country tends to make the country unattractive due to security threats, and most tourists are likely to avoid such countries. On the other hand, countries that are politically stable are more than likely to attract more tourists.

Legal factors: Decisions made by governments to either open or restrict their borders have an impact on the economic growth of the tourism industry. There are countries that impose difficult conditions in a bid to restrict the flow of foreigners into their country, which in turn affects the number of tourists visiting those countries. Countries that operate on open borders have a higher flow of tourists than countries that have strict requirements that need to be met prior to entering their territory.

Ethical factors: Ethics refers to the various principles, values and rules affecting the engagement of person-to-personal interactions and behaviours that are intrapersonal in nature (Kiarie, 2022). Ethical tourism has been emphasized in recent times, and tourists are more conscious of the impact that their travel activities have on their travel destinations. Exploitation of children and women, sex tourism, dark tourism and animal welfare are some of the ethical dilemmas facing the tourism industry (Sonia, 2014).

In summary, though each component in the STEEPLE framework is distinctive, as discussed above, all the components come into play when understanding the influence of the external environment on the activities of women entrepreneurs. For example, the use of resources in a manner that enables economic, social and aesthetic needs to be met while still maintaining the cultural integrity requires technical support systems (Roe & Khanya, 2001). About empowerment, studies on Pro-poor Tourism (PPT) reveal how increased net benefits for the poor can ensure that poverty reduction is a priority towards achieving economic prosperity for the target group (Roe & Khanya, 2001). This is related to the economic and social factors. In Kenya, tourism extends to the local communities in a bid to reduce the poverty levels. The geographical distribution of the tourism landscape in Kenya is twofold, i.e. wildlife safaris and beach holidays. However, in recent times, a third component has evolved, i.e. cultural tourism. The Maasai community in Kenya is renowned for its rich cultural heritage. However, the general living standards of the community are below average in comparison to other ethnic communities. Over the last two decades, the Maasai community has come up with Maasai cultural tourism (Ondicho, 2017).

Methodology

The focus of this chapter is on the women entrepreneurs in the Maasai community who operate micro and small enterprises in Kajiado County, Kenya, and their active engagement in the local tourism industry through production and sale of products for the local and foreign tourists. A trip to Kenya would be incomplete without carrying back some Maasai culture memorabilia, and most of the memorabilia are handcrafted by women ranging from beads, shawls, necklaces, earrings, and garments, among other items. The Maasai cultural tourism involves giving tourists an experience of the Maasai way of life right from their handcrafted products to their homesteads as well as the vast grasslands where the community grazes its livestock.

Qualitative data were utilized, and data collection was done through interviews. A total of ten women entrepreneurs from the Maasai community who run micro and small enterprises selling various products targeted at the local and foreign tourists were interviewed. The women entrepreneurs were from three localities/villages where the Maasai community resides, namely, Kisamis Town, Kiserian Town and Ngong Town in Kenya. The use of interviews was selected due to the low literacy levels among the women in the target group and due to the need to translate their responses from the local dialect (Maasai language) and Swahili into English.

The results of the interviews were analysed using inductive method of data analysis focusing on thematic content analysis. A theme refers to a patterned meaning or response obtained from the data resulting from the research question (Braun & Clarke, 2012). Thematic analysis involves identification of patterns or themes that occur when collecting qualitative data (Maguire & Delahunt, 2017). The choice of using thematic analysis is pegged on the ability to analyse data using a data-driven or theory-driven approach, the flexibility of applying it in a broad aspect of epistemological and theoretical frameworks and the possibility of applying it on various sample sizes, study questions, volumes of data and designs (Kiger & Varpio, 2020). Analysis of the qualitative data collected was done using six steps, namely, annotation of the transcripts, aligning the data with key themes, connecting the categories, positioning the categories, undertaking a deep data segment analysis and compilation of the findings (Maguire & Delahunt, 2017).

A total of 13 general and specific interview questions were utilized to collect data from the respondents. The responses were recorded and categorized into the following themes: products manufactured, sources of raw materials, buyers of the products, export sales, source of funding, average income, government support, non-governmental organization's support, personal benefits obtained from their businesses, beneficiaries of their businesses, involvement of men in the trade and challenges experienced while operating their businesses. Table 6.1 provides a tabulation of the interview question, the response received and the corresponding themes. The responses were received using Swahili language with a mixture of Maasai language. These responses were thereafter translated into English.

Table 6.1 Interview Questions and Responses That Were Received

No	*Interview Question*	*Response*	*Theme*
1	What products are manufactured by the local communities?	Earrings, belts, kaskas, necklace, bags, ngiso (keyholders), bangles, bracelets, olkerepa (head ornament), lessos and shukas(shawls), telling tales, narrating history of the community, singing traditional songs for a fee	Products manufactured
2	What are the sources of raw materials for their products and what raw materials are used?	Thread, recycled wire, beads, skin for the belts, manilla, nyororo (metallic chains), recycled metal, wood	Sources of raw materials
3	Who are the main buyers of the products?	Local tourists, own consumption, sale to residents of local areas within Kajiado County, international tourists in Mombasa	Buyers of the products
4	Are the main buyers of their products foreign or local buyers?	Both foreign and local buyers	Buyers of the products
5	Are there any products that are exported?	There are no exports. However, they have formed chamas (women's self-help groups) to assist them in getting to the export level	Export sales

(*Continued*)

Table 6.1 (Continued)

No	*Interview Question*	*Response*	*Theme*
6	Where do the businesses get their capital?	From their savings, loans from saccos, chamas (women's self-help groups), support from their children, from livestock, milk sold, barter trade within the community	Source of funding
7	What is the average income from the products sold monthly?	An average of Kshs 5,000 per month ($39.32)	Average income
8	Does the government give the businesses direct or indirect support?	There is no support from the government	Government support
9	Are there NGOs that support the businesses?	There is no support from NGOs	Non-governmental organization support
10	How have their businesses assisted them at a personal level?	Exposure in the marketplace to more products, networking in the marketplace leading to formation of chamas (women's self-help group), learning the Swahili language, visiting foreign countries (one of the women entrepreneurs went to the USA and her photo was used on the billboards but she was not paid)	Personal benefits obtained from their businesses
11	How many direct and indirect people rely on their businesses to meet their daily needs (nuclear family, extended family, etc?)	An average of ten dependents	Beneficiaries of their businesses
12	Are there men who are involved in the trade?	No. It was rare for men to be involved unless in the production process, e.g. preparation of animal hides	Involvement of men in the trade
13	What are the main challenges that face their cottage industries?	Lack of capital, difficulty in getting buyers, difficulty in obtaining county licences, unscrupulous people duping them on finding markets for their products, difficulty in accessing loans, walking long distances from home to the markets, locals don't want to buy at good prices, lack of marketing skills, lack of mobile phones and internet connectivity to market their products	Challenges experienced while operating their cottage industries

Source: Prepared by the Author.

Figure 6.1 Products manufactured by women from the Maasai Community

For interview question number one, some of the products that are manufactured and sold by the women have been provided in Figure 6.1.

Findings

The research focused on uncovering the factors that influence the empowerment of women entrepreneurs in the tourism industry in Kenya and examine how social policies can enable women entrepreneurs to improve their livelihoods and those of the tourism communities through entrepreneurship. Below is an analysis of the finding based on the objectives of the study.

Factors that Influence the Empowerment of Women Tourism Entrepreneurs

Based on the STEEPLE framework, below are the findings of the factors that influence empowerment of women entrepreneurs operating cottage industries in Kenya.

Social Factors

The findings showed that the women entrepreneurs were dependent upon the family and social set up to raise capital for their businesses. In addition, the tourism enterprises operated by the women entrepreneurs were supporting an average of ten dependents both directly and indirectly, which illustrated that the women entrepreneurs were central to meeting the daily needs of their dependents. The absence of men in the cottage industries operated by the women was evident and being a nomadic community, and this is a pointer to stereotyping related to these cottage industries being run by women. The women entrepreneurs also highlighted various personal benefits that were derived from their businesses with the prominent ones being getting more exposure in the marketplace, learning Swahili language and visiting foreign countries. One of the respondents mentioned that she visited the USA where she prominently featured on billboards, but she did not derive any substantial economic benefit. With the low income from the cottage industries and the high number of dependents, growth of their small and micro enterprises would be impacted since the little profit that was made was not ploughed back into the business and was instead used to take care of their dependents. The impacts of a social and cultural nature are inevitable, and they need to be assessed to avoid compromising the sustainable development strategy in the tourism industry (Baum & Ndiuini, 2020).

Technological Factors

The cottage industries operated by the women in the Maasai community produce various products such as ornaments, artefacts and garments. It was noted that the products were handmade and there was no use of technology in the production process. From the interviews, the women also earned a living by telling tales, narrating the history of the community and singing traditional songs for both local and international tourists who visited them. It was evident the women produced a broad spectrum of ornaments, artefacts and garments. From the findings, there was evidence that the micro and small enterprises were not leveraging on technology to market their products. The cost of smart phones has reduced over the years, and various product and tourist marketing sites exist. In addition, with multiple delivery options in place, the consumers do not have to be physically present to purchase their products. Technology simplifies industries, and the linking of these micro and small enterprises with technology, marketing sites and online delivery channels can increase the earnings from their products. Businesses can spur their growth by utilizing technology to communicate with potential customers, and due to the diverse channels of interactive media, consumers are likely to spot more product information (Jain & Yadav, 2017).

Economic Factors

The buyers of the items produced were both local and foreign tourists. However, for the foreign tourists, their products were transported to the coastal city of Mombasa since this is where majority of the tourists visit frequently. There were no products that were being exported by the women entrepreneurs. However, there were efforts through their chamas (women's self-help groups) to make the export of products a reality.

Most of the sources of funding were obtained informally, from their personal savings, loans from their chamas (women's self-help groups), support from family and friends. Barter trade within the local community was also used whereby livestock was traded for the products manufactured. The average income from the products sold was Kshs 5,000 or $39.32 per month. This income was solely derived from the sale of the ornaments, artefacts and garments. The sources of funding that were mentioned were not adequate to assist the women in growing and sustaining their businesses. The women were locked out from the mainstream financing systems that were available, and this was attributed by the fact that they were not deemed financially desirable by various potential financing organizations as well as their low literacy levels. Access to capital was uncovered as a major hinderance in the growth of the small and micro enterprises. As such, the women entrepreneurs need to explore opportunities for value addition to their products, which could yield higher profits. It's worth considering that tourism has the potential of providing an economic base for communities that solely depend on cultural and natural resources (Christie, Fernandes, Messerli, & Ward, 2014).

Environmental Factors

Most of the raw materials for production of jewellery were sourced from nature and these materials are considered renewable either through biodegradable processes or through agricultural processes which eventually balance concerns regarding the environment (Ray, 2019). From the research findings, the raw materials utilized to produce the various crafts by the micro and small enterprises were environmentally friendly products. A key lesson from these businesses is the use of sustainable raw materials to produce environmentally friendly products, and research shows that particular products that are associated with sustainable raw materials result in doubling the growth in comparison to other products that are not made from sustainable raw materials (White, Hardisty, & Habib, 2019). The activities of the women entrepreneurs in the Maasai community have tried to achieve sustainable tourism, which refers to making a commitment to reduce the impact of tourism activities on the environment and preserve the local culture (Van der Bank, 2016). These micro and small enterprises can leverage on their products being manufactured using sustainable raw materials as a marketing strategy to the current generation of consumers who are keen on purchasing environmentally friendly products. By exploiting the aspect of their products being environmentally friendly, it was possible for them to charge a premium especially when selling to consumers who are conscious of the environment.

Political Factors

The findings noted that none of the women engaged in the business were receiving government support towards marketing and selling their products. The micro and small enterprises were also not able to access capital to grow and sustain their businesses. A key finding was the failure by the government to involve itself in supporting these businesses. While the government may not fund the business, they can put in place policies and structures to promote the growth of these businesses. Tourism is a visible sector with a potential for high earnings that can increase a country's GDP and employment opportunities (Khan, Hassan, Fahad, & Naushad, 2020). Therefore, it would be prudent and politically worthwhile for the government to offer support and incentives for these women entrepreneurs. The government can create a platform that is specifically tailored to meet the funding needs of these enterprises. To this end, the government of Kenya has introduced a fund known as the Hustlers fund which has a start-up capital of $380,000,000 that allows Kenyan citizens to borrow money through their mobile phones (Hustlers Fund, 2023). A portion of these funds can be allocated towards advancement of social policies geared to support women in the tourism industry. The women owning these micro and small enterprises need to be sensitized on how to access these funds to grow their business.

Legal Factors

There was no legislation that was specifically targeted at these women entrepreneurs and their businesses. In addition, they were not aware of legislation that was focused on their micro and small enterprises. Registration of businesses in Kenya has been simplified using technology in the Business Registration Service website and with the limited access to technology faced by the women entrepreneurs, and this could pose a challenge for them as they seek to formalize their businesses. Despite being micro and small enterprises, they were not exempt from paying market licence fees to the County Government. The market licence fees are required for them to sell their wares in the marketplace designated by the County Government. Considering the meagre earnings that the women entrepreneurs were making, an exemption or reduction of the market licence fee can reduce their cost of business.

Ethical Factors

Various individuals and organizations were interested in aiding the women running the micro and small enterprises. However, from the findings, it was clear that some individuals/organizations were taking advantage of these women. Both the suppliers and consumers in the tourism industry are still unclear on differentiating between what is right and wrong with regard to the tourism industry, and this can be solved through education for both parties and development of a universal set of codes relating to ethics in the tourism industry (Khan, 2014). A structured approach for aiding these micro and small enterprises needs to be formulated. The chamas (self-help groups) formed by these women can be used as avenues to facilitate receiving of assistance from those individuals/organizations who might be willing to help.

The Role of Social Policies in Enabling Women's Tourism Entrepreneurship

When the society empowers its women, a wider impact is felt at the family level, community setup and national level. The impact of self-help groups in raising capital, referred to as "Chamas", was evident in this study. The self-help groups formed by the women to support their businesses also indicates that the enterprises were valued towards sustaining their livelihood. Through these self-help groups, women were able to save their earnings and at the same time obtain loans to fund their businesses. Studies have shown a positive association between self-help groups and increased business income, adoption of technology as well as improved productivity and market involvement (Biscaye et al., 2014).

Targeted interventions meant to assist the women running these micro and small enterprises to achieve improved economic and social well-being that needs to be formulated. The Government of Kenya has established two agencies that are targeted at assisting micro and small enterprises and women entrepreneurs, namely, the Women Enterprise Fund (WEF) and the Micro and Small Enterprises Authority (MSEA). However, little is known about these two agencies by the women entrepreneurs who were the subject of this study. The Women Enterprise Fund (WEF) was established in 2007, and it assists women self-help groups in accessing alternative financial services (Kenya Vision 2030, n.d.). The Micro and Small Enterprises Authority (MSEA) is mandated with the formulation and review of policies and programmes for micro and small enterprises (Micro and Small Enterprises Authority, 2020). From a review of the two agencies, their policies were not specifically addressing the role played by women entrepreneurs in the tourism cottage industries. In addition, the women entrepreneurs in this study were not aware of existence of these organizations and the existence of such alternative sources of funding. In summary, the government and non-governmental organizations should be actively involved towards supporting these micro and small enterprises to achieve economic growth through offering education, financing and marketing avenues for their products.

The women entrepreneurs in this study were optimistic about their businesses and determined to operate their businesses despite the challenges they were facing. From the findings, the women entrepreneurs were also actively involved in taking care of their families while still running their enterprises. At the same time, they were receiving minimal assistance from men in the community towards improving their cottage industries. Research studies suggest that the presence of a male role model is critical in determining the possibility of a woman entrepreneur venturing into sectors that are traditionally male-dominated and associate with higher profits (Diop, 2017).

Discussion

The STEEPLE model used in this study provided an understanding of the external factors that influence social policies for supporting women in the tourism industry in Kenya. The challenges facing women entrepreneurs can be grouped into business and societal challenges. Business-related challenges include lack of capital, difficulty in getting buyers, difficulties in obtaining business/trade licences and permits, local tourists paying low prices for their products, lack of avenues for marketing their products and lack of technology use in the production and marketing of their products. To note is that the micro and small enterprises owned by the women produced a wide array of products. While the handmade items were crafted with precision and skill, the enterprises could benefit from technology in the production of the products.

The use of sustainable raw materials is a selling point for their products since a good proportion of the raw materials utilized was environmentally friendly. However, at the selling point, the women were not mentioning the fact that the raw materials used in production of the items were environmentally friendly. The potential of exporting their products was not being exploited. While the women mentioned that they had formed chamas (self-help groups) to assist in the efforts, they needed to rope in more stakeholders to assist them in, for example, actively seeking the help of the government and non-governmental organizations. The businesses operated by the women had the potential to generate more revenues; however, the current earnings of an average of $39.32 per month were not adequate. Ethical issues were uncovered whereby while foreigners were interested in these cottage industries and showcasing them in their country, the local women were not deriving any economic benefits. The businesses operated by the women supported several dependents. However, comparing the average income versus the number of dependents, their businesses were not adequate in meeting their needs.

The women also faced challenges that are social in nature and gendered. For instance, the women spoke about walking long distances from their homes to the marketplaces. Traditionally, the Maasai are a nomadic community relying on livestock as the main source of livelihood. However, with the effects of climate change, it has been difficult for the community to sustain nomadism. As such, women have ended up being providers for the family through the sale of artefacts and products, which places additional burden on them apart from being caregivers. From the findings and the challenges that were collected, it was noted that there was very little progress that was being made towards coming up with social policies to support women in the tourism industry in Kenya. Some of the challenges could be resolved by the involvement of the government, for example, the government could make it easier for these businesses to obtain business/trade licences and permits. In addition, various actors including the government can provide marketing avenues for these cottage industries to enable them to derive more benefits from selling products to tourists.

The existing political climate has the potential of exposing the tourism sector to volatility and uncertainty (Mayaka & Prasad, 2012). This chapter uncovers how the cottage industries run by women and the men play limited roles in the business activities in these industries. Social factors cover demographics, religion, society values and the need to accept social differences and perform actions necessary to ensure that there is preservation of stability within society to ensure the development of the tourism industry (Nicolae, 2018). The use of technology in advertising and media was a key component in marketing and influencing the tourists' knowledge (Ikiara, 2001). The limited use of technology in marketing the products was acknowledged as an obstacle, and financial costs were a major hinderance to the adaptation of technology by the women entrepreneurs in their businesses; environmental factors explored in other studies focused on the preservation of the environment, use of recycled products and education of the communities on environmental conservation (Nicolae, 2018). This study uncovered the use of recycled products in the production of the ornaments and artefacts.

This study has shown that women entrepreneurs utilize their earnings to enhance their personal welfare and to support both direct and indirect dependents. Previous research has shown that inexpensive solutions and minor changes in the support offered to female entrepreneurs have a major impact on the potential of increasing their earnings (Gaye, 2018). The findings add to the existing research that shows that empowerment of women leads to broader development impacts through generation of income and creation of jobs (Kimbu & Ngoasong, 2016).

In the part of the tourism industry, the artefacts sold by the women entrepreneurs are highly priced by tourists since these are products that serve a functional purpose and are also aesthetically appealing, and the earnings from the sale of these products add up to the total earnings gained by the tourism industry in Kenya. This illustrates that the Maasai women entrepreneurs also contribute to the earnings and success of the tourism sector.

Conclusion

From the study, technology plays a critical role in business development. In this study relating to women entrepreneurs in the tourism industry, technology is critical to the registration of their businesses, production process for the artefacts and marketing of the products. A key recommendation for the women entrepreneurs is for them to embrace and invest in technology to increase the income from their businesses. For the government, the support provided to these women entrepreneurs through enabling them to access technology, assisting in marketing their products and making it easy for them to do their business would result in creation of employment and eventually contribute to the tax that the government collects.

Note

1 $1,182,973,614.

References

Baum, T., & Ndiuini, A. (2020). Tourism in Africa: Issues and prospects. In Baum, T. & Ndiuini, A. (eds.) *Sustainable Human Resource Management in Tourism* (pp. 9–27). Cham: Springer.

Biscaye, P., True, Z., Clark, C., Harris, K. P., Anderson, L., & Gugerty, M. K. (2014, December 5). *Evans School of Public Affairs.* Retrieved from University of Washington: https://epar.evans.uw.edu/sites/default/files/epar_283_shg_evidence_review_brief_10.23.20.pdf

Braun, V., & Clarke, V. (2012). Thematic analysis. In Cooper, H., Camic, P. M., Long, L., Panter, A.T., Rindskopf, D. & Sher, K.J. (eds.) *APA Handbook of Research Methods in Psychology* (pp. 55–71). Washington, DC: American Psychological Association.

Bush, T. (2019, November 17). *Pestle Analysis of the Tourism Industry.* Retrieved from Pestle Analysis: SWOT and Business Analysis Tools: https://pestleanalysis.com/pestle-analysis-of-tourism/

Christie, I., Fernandes, E., Messerli, H., & Ward, T. L. (2014). *Tourism in Africa: Harnessing Tourism for Growth and Improved Livelihoods.* Washington, DC: The World Bank.

Deen, A., Harilal, V., & Achu, F. (2016). Women in tourism: Experiences and challenges faced by owners of small accommodation establishments. *African Journal of Hospitality, Tourism and Leisure*, *5*(3), 1–15.

Diop, M. (2017, March 08). *Unleashing the Potential of Women Entrepreneurs in Africa.* Retrieved from World Bank Blogs: https://blogs.worldbank.org/nasikiliza/unleashing-the-potential-of-women-entrepreneurs-in-africa

Gaye, D. (2018, November 29). *Female Entrepreneurs: The Future of the African Continent.* Retrieved from The World Bank: https://www.worldbank.org/en/news/opinion/2018/11/29/women-entrepreneurs-the-future-of-africa

Gendernet. (2011). *Women's Economic Empowerment.* Paris: OECD.

Hustlers Fund. (2023, March 2). *Hustlers Fund.* Retrieved from Hustlers Fund: https://www.hustlerfund.go.ke/

Ikiara, M. M. (2001). *Policy Framework of Kenya's Tourism Sector since Independence and Emerging Policy Concerns.* Retrieved from Kenya Institute for Public Policy and Research Analysis: https://repository.kippra.or.ke/bitstream/handle/123456789/2736/WP2.pdf?sequence=1&isAllowed=y

ILO. (2002). *Women and Men in the Informal Economy.* Geneva: International Labour Organization.

Jain, E., & Yadav, A. (2017). Marketing and technology: Role of technology in modern marketing. *Journal of Business and Management,* 49–53.

Kenya Vision 2030. (n.d.). *Women Enterprise Fund.* Retrieved from Kenya Vision 2030: https://vision2030.go.ke/project/women-enterprise-fund/

Khan, N., Hassan, A. U., Fahad, S., & Naushad, M. (2020, March 24). *Factors Affecting Tourism Industry and Its Impacts on Global Economy of the World.* Retrieved from Social Science Research Network: https://papers.ssrn.com/sol3/papers.cfm?abstract_id=3559353

Khan, S. (2014). Tourism Business: An Ethical Dimension. *International Journal of Hospitality & Tourism Systems*, 63–70.

Kiarie, R. K. (2022). *Ethical Marketing: Concepts and Reasons.* Bingley: Emerald Publishing Ltd.

Kiger, M., & Varpio, L. (2020). Thematic analysis of qualitative data: AMEE Guide No. 131. *Medical Teacher,* 1–9.

Kimbu, A. N., & Ngoasong, M. Z. (2016). Women as vectors of social entrepreneurship. *Annals of Tourism Research, 60*, 63–79.

Maguire, M., & Delahunt, B. (2017). Doing a thematic analysis: A practical, step-by-step guide for learning and teaching scholars. *All Ireland Journal of Teaching and Learning in Higher Education (AISHE-J)*, *9*, 3351.

Mayaka, M. A., & Prasad, H. (2012). Tourism in Kenya: An analysis of strategic issues and challenges. *Toruism Management Perspectives*, 48–56.

Micro and Small Enterprises Authority. (2020). *Kenya Micro and Small Enterprises Policy for Promoting Micro and Small Enterprises (MSEs) for Wealth and Employment Creation.* Retrieved from Micro and Small Enterprises Authority: https://msea.go.ke/wp-content/uploads/2021/07/MSEs-Policy.pdf

Nicolae, P. (2018). Use of the PESTEL model in the management of the tourism branch of the Republic of Moldovia. *Ovidius University Annals, Economic Sciences Series*, *0*(1), 370–375.

Ondicho, T. (2017). The contribution of tourism to poverty reduction: A case study of communities living adjacent to Amboseli National Park, Kenya. *International Journal of Development Dialogue*, *2*(1), 34–53.

Ray, S. (2019, April). *New Trend in Jewelry Industry and Sustainable Materials to Develop Lifestyle Pproducts.* Retrieved from ResearchGate: https://www.researchgate.net/publication/332672275_New_trend_in_Jewelry_industry_and_Sustainable_materials_to_develop_lifestyle_products

Revfine. (2021). *13 Key Technology Trends Emerging in the Travel & Tourism Industry.* Retrieved from Revfine: https://www.revfine.com/technology-trends-travel-industry/

Roe, D. and Urquhart, P. (2004) Pro-poor tourism: Harnessing the world's largest indus-try for the world's poor; turning the rhetoric into action for sustainable developmentand poverty reduction. In Bigg, T. (ed.) Survival for a Small Planet – The Sustainable Development Agenda (pp. 309–325). London: Earthscan

Saidova, M., & Muminjonova, E. K. (2022). Using steeple analysis in business process (in case of Ubzekistan). *International Scientific and Current Research Conferences*, New Jersey, 48–60.

Santander Fundación. (2020, October 15). *Entrepreneurs in Santander BEST Africa.* Retrieved from Santander Fundación: https://www.fundacionbancosantander.com/en/social-action/santander-best-africa/women-entrepreneurs/leontine-keita

Seyitoglu, F., & Ivanov, S. (2020). A conceptual framework of the service delivery system design for hospitality firms in the (post-) viral world: The role of service robots. *International Journal of Hospitality Management, 91*, 102661–102661.

Sonia, K. (2014). Tourism business: An ethical dimension. *International Journal of Hospitality & Tourism Systems, 7*(2), 63–70.

Statista. (2015, September 30). *Percentage of GDP Contributed by Women 2015, by Region.* Retrieved from www.statista.com: https://www.statista.com/statistics/523838/women-share-of-gdp-region/

Statista. (2022, August 1). *Tourism Earnings from International Arrivals in Kenya 2005–2021.* Retrieved from Statista: https://www.statista.com/statistics/1140037/tourism-earnings-from-international-arrivals-in-kenya/

Swiss Agency for Development and Corporation (SDC). (2003). *Gender Equality: A Key for Poverty Alleviation and Sustainable Development.* Bern: Swiss Agency for Development and Cooperation (SDC).

Tugjamba, N., Walkerden, G., & Miller, F. (2023). Adapting Nomadic pastoralism to climate change. *Climatic Change, 176*, 28.

UN Women. (2020, December 1). *Advancing Women's Economic Empowerment in the Tourism Sector in COVID-19 Response and Recovery.* Retrieved from https://www.unwomen.org/sites/default/files/2022-05/Advancing-womens-economic-empowerment-in-the-tourism-sector-in-COVID-19-en.pdf

UN Women. (n.d.). *Economic Empowerment of Women.* New York: UN Women. Retrieved from UN Women from www.unwomen.org/sites/default/files/Headquarters/Attachments/Sections/Library/Publications/2013/12/UN%20Women_EE-Thematic-Brief_US-web%20pdf.pdf

USAID, & Scott, L. (2021). *A Briefer: Gender Inequality Causes Poverty.* USAID. Retrieved from https://banyanglobal.com/wp-content/uploads/2021/03/Gender-Inequality-Causes-Poverty-Briefer.pdf

Van der Bank, M. C. (2016). A legal framework for sustainable tourism development in areas affected by natural and manmade disasters in South Africa. *African Journal of Hospitality, Tourism and Leisure, 5*(2), 1–12.

Ward, L. T. (2010). *Global Report on Women in Tourism 2010.* Madrid: World Tourism Organization and UN Women.

White, K., Hardisty, D. J., & Habib, R. (2019). *The Elusive Green Consumer.* Retrieved from Harvard Business Review: https://hbr.org/2019/07/the-elusive-green-consumer
World Bank. (2007). *Global Monitoring Report: Confronting the Challenges of Gender Equality and Fragile States.* Washington, DC: World Bank.
Young, L. (2012). *Challenges at the Intersection of Gender and Ethnic Identity in Kenya.* Minority Rights Group International. Retrieved from https://minorityrights.org/app/uploads/2023/12/download-1164-challenges-at-the-intersection-of-gender-and-ethnic-identity-in-kenya.pdf

7 The Gendered Dynamics of Financing Tourism Entrepreneurship

Anna de Jong, Michael Zisuh Ngoasong and Albert Nsom Kimbu

Introduction

Within entrepreneurship scholarship, access to finance is often narrated as a challenge that effects women entrepreneurs, where it is understood that women-owned businesses have the lowest success rate in accessing both formal and informal finance (Kim 2006; Ngoasong and Kimbu 2016). There are several hypotheses as to why this may be the case, including the well-rehearsed arguments that women are reluctant borrowers, demand lower levels of finance and are less likely to engage with angel and equity finance (Marlow 2013) and patriarchal systems which still tend to favour men (Ngoasong and Kimbu 2019). Much of this research narrates women as subordinate, where there is an inherent assumption that there is something associated with the performance of women that limits their ability to command entrepreneurial performance, in relation to entrepreneurial men (Marlow and Swail 2014).

Such constructions are inhibiting for two central reasons. First, taking focus with the limitations of women entrepreneurs fails to acknowledge the structural governance and institutional inequalities that constrain women's (and many others) abilities to enact entrepreneurial objectives. For example, women's enterprises are disproportionally predominant within the services sector. They are less likely to offer opportunity for higher value-added ventures and are considered riskier by funders, who are in consequence likely to charge higher interest rates to counter risks (Marlow and Patton 2005), possessing implications for growth prospects. Whilst as women transition from employment to entrepreneurship, they are less likely to have been in employment positions that would facilitate accumulation of social and economic capital that is valuable for new firm creation (Marlow and Swail 2014). Second, contemporary feminist perspectives caution that consistently taking focus with women entrepreneurs obscures the finding that *most* entrepreneurs are reluctant borrowers who aim to minimise their financial liabilities, counter to normative narratives that business owners desire to grow firms as quickly as possible.

Feminist entrepreneurship scholars thus argue that by shifting focus away from that of 'women' as a homogenous group (in comparison to 'men'), to a more nuanced assessment of all genders, allows us to better understand similarities and

DOI: 10.4324/9781003315858-7

overlaps between and within gender categorisations, alongside variations, as well as how political and social structures produce gendered outcomes (Ahl 2004; Hill, Leitch, and Harrison 2006). Thus, we here respond to Hill et al.'s (2006) claim that a more heterogenous approach to the study of gender, entrepreneurship, and finance is necessary to increase the understanding of gendered outcomes in accessing finance, a claim that we posit still holds, despite first being shared 16 years ago.

In recognition, this chapter aims to undertake a heterogenous approach to the study of gender, entrepreneurship, and finance, to examine how gender intersects with place-based specificities to influence perceptions and motivations regarding finance. We apply a feminist geographic lens, to capture the place-specific elements of governance, policy, society, and culture, that intersect with gender to influence motivations and perceptions. Ghana and Nigeria are taken as the two case study contexts, chosen as they offer opportunity to look beyond the dominance of Global North perspectives within the literature (Figueroa-Domecq, de Jong, and Williams 2020). This chapter focuses on the service sector because of the high proportion of women's entrepreneurship, compared to manufacturing, and the specific challenges in accessing finance associated with the service sector of the economy.

Literature Review

Entrepreneurship is positioned as a means through which to achieve high economic growth and job creation (Doran, McCarthy, and O'Connor 2018). As such, governments and policy makers are often keen to establish rules and regulations that enable an environment that facilitates business development – with such rules and regulations perceived to be a key pillar in determining the success of entrepreneurial activities within a geographic region (Abubakar 2015). Newly established micro-, small- and medium-sized enterprises (MSMEs) are perceived as particularly risky by funders, leading to barriers for nascent entrepreneurs seeking funding (Marlow and Patton 2005). Policy support is considered pivotal to constructing an environment conducive to entrepreneurship because a lack of access to funding in the early stages of business development impedes the ability of businesses to establish, innovate and grow to their full potential longer term – countering policy intentions to utilise entrepreneurship as a vehicle through which to generate economic growth and job creation.

And yet, despite such intentions, government and policy support are often considered weak or ineffective, as is the case in Ghana and Nigeria (Udefuna and Uzodinma 2017), whereby in these regions it can be challenging to access formal finance to undertake entrepreneurship as government institutions struggle to create enabling environments to assist active entrepreneurs (Ngoasong and Kimbu 2019). As such, according to the World Bank's Enterprise Surveys, 45% of firms in Africa identified access to finance as a key obstacle to their business, compared to 13% in OECD countries. Across Africa access to formal finance is gendered, whereby women entrepreneurs receive a small fraction of the total capital available for SME investment (Equal Footings Report 2014). This differential is considered one of the main reasons women dominate in running nascent and new businesses, and men dominate in running businesses that are more than three and a half years old (GEM 2010). Challenges regarding access to finance prevail despite Nigeria having introduced several policies seeking to promote entrepreneurship since the 1970s (Udefuna and Uzodinma 2017), whilst in Ghana the first attempt to support business was introduced in 1985, through the National Board for Small-Scale Industries (Obeng and Blundel 2015). Recently, there have been increasing calls for more specific policy initiatives intended to encourage women's entrepreneurship to counter gendered inequalities (Huysentruyt 2014); such calls though are yet to be actively responded to by the Ghanaian and Nigerian governments.

Despite the integral role of governance and policy in facilitating access to formal finance, we know little about the impact of governance and policy on women entrepreneurs and their access to finance (Leitch, Welter, and Henry 2018). This is a surprising omission within the entrepreneurship literature given the long-held construction within this scholarship that women are less effective in obtaining formal finance for their entrepreneurial endeavours. Yet an omission that perhaps prevails because it is often women who are viewed to be at fault for these disparities, rather than the existence of structural constraints (Leitch et al. 2018).

In this chapter, we engage with Marlow and Swail's (2014) critical critique of such constructions, suggesting that structural influences position women in socio-economic spaces that militate against entrepreneurial endeavour and that these structural influences are erroneously and stereotypically interpreted as agentic shortcomings (Marlow and Swail 2014). Such problematic constructions have become dominate within the literature because whilst there has been an enhanced focus on 'women' as a category of analysis, this engagement has been conducted without critically attuning to the role of structure. Much scholarship in consequence acknowledges gender as an issue but maps the construct onto women and their behaviours, a process that articulates and upholds the gender hierarchy through a focus upon individual women and their lack of entrepreneurial skills and attitudes (Marlow and Swail 2014). This is a construction that fails to acknowledge the structural assumptions which constrain activities and inform motivations and perceptions within specific contexts (Marlow and Swail 2014).

Such constructions, in part, prevail because research has tended to be dominated by scholars in the US, UK and Europe, who have prioritised certain ideas, methods and frameworks – producing positivist qualities defined through conceptions of essentialism and objectivity (Leitch et al. 2018). Progress has thus been slow in understanding the complexity of gendered experiences across different contexts, ensuring that problematic conceptions of women entrepreneurs and their inability to access finance have remained upheld. To overcome such dichotomous gendered thinking, Leitch et al. (2018) suggest that more research ought to take focus on the everyday ways entrepreneurs draw on innovative techniques to finance their enterprises in different contexts – an approach to research that would anchor entrepreneurial experiences in place to explore how the social and political contexts inform motivations and perceptions of finance. They further note the value in understanding the specific challenges facing women in the Global South in accessing formal and informal finance. Whilst extant literature has taken focus on women's agentic shortcomings, what has remained less visible are the deeply rooted stereotypical thinking and attitudes prevalent within different cultures and within governance structures (Kwapisz and Hechavarría 2018).

This chapter therefore employs a feminist geographic approach and engages with the empirical contexts of Ghana and Nigeria, to gain insights into the relationships between policy, entrepreneurship, finance and gender. Feminist geography recognises the importance of place in understanding gendered performance, cultural and social norms and policy constructions whilst remaining attuned to social differences (including gender, age, race and so on). In taking this approach we identify insights into entrepreneurs' perceptions and motivations of finance, contributing insights into the complex ways gender intersects with place to produce disparities in entrepreneurial access to finance.

Method

Ghana and Nigeria are valuable empirical contexts for this research because they are under-researched tourism destinations, with high numbers of individuals engaging with entrepreneurship. In Ghana, MSMEs employ more than 80% of the workforce and generate 70% of national output (Ministry of Trade and Industry, Ghana 2019). In Nigeria, SMEs account for 96% of businesses, 84% of employment and 48% of national GDP (PwC 2022). The numbers of women entrepreneurs are also high in both Nigeria and Ghana, providing unique contexts where entrepreneurship is close to gender parity, rather than being male dominated as is often the norm (Meunier, Fantoni and Kouhlani-Nolla 2022). In Ghana, for example, it is estimated that 46% of businesses are owned and operated by women, primarily within agriculture, retailing and service provision (Mastercard 2018). Whilst 41% of Nigerian women are entrepreneurs owning/managing a range of MSMEs (Bill and Melinda Gates Foundation 2020). Together, Ghana and Nigeria represent a significant number of women entrepreneurs within the geographic region of West Africa, a region that has experienced similar structural challenges and forms of governance within the context of entrepreneurship (Sackeyfio 2021).

In terms of the method, semi-structured interviews were conducted with 9 entrepreneurs in Accra and Cape Coast Ghana (two women and seven men) and 18 entrepreneurs in Lagos, Nigeria (12 women, 6 men), in March 2019. Entrepreneurs operated a range of tourism businesses, varying in size (micro, small and medium), operational age (nascent to established) and service provision type (including, for example, accommodation, tour operators, caterers, transportation, food, beverage and so on) (participant attributes are outlined in Table 7.1). Interviews lasted around one hour and were recorded, following consent from the participant. All participants were given a pseudonym.

Recruitment took place through purposive snowball sampling, where care was taken to select participants from a range of genders and enterprise types, sizes and ages. This approach ensured that a range of perspectives and motivations relating to finance were identified. Recruitment commenced through engagement with local

Table 7.1 Participant Attributes

Participant	*Location*	*Gender*	*Age Group*	*Business*	*Marital/Family Status*
Abeo	Nigeria	Man	40–49	Hotel Management	Married, children
Adaku	Nigeria	Woman	20–29	Food and Beverage	Single
Adom	Ghana	Man	30–39	Hospitality Management	Married
Aisha	Nigeria	Woman	30–39	Hotel Management	Single
Amaa	Ghana	Man	50–59	Regional Development Agency	Married, children
Anto	Ghana	Man	30–39	Hotel Management	Married, child
Buki	Nigeria	Woman	30–39	Youth Development Initiative	Married
Chi	Nigeria	Man	50–59	Tour Operator	Married, children
Chisom	Nigeria	Woman	40–49	Tourism Media	Married, children
Dede	Ghana	Man	30–39	Taxi Business	Married
Ebo	Ghana	Man	40–49	Regional Tourism Authority	Married, children
Ezinne	Nigeria	Woman	30–39	Event Management & Catering	Single
Fola	Nigeria	Woman	40–49	Ministry of Arts & Culture	Married, children
Ikemba	Nigeria	Man	40–49	Tourism Media and Marketing	Married, children
Ileara	Nigeria	Woman	30–39	Hospitality Catering	Single
Isoken	Nigeria	Woman	50–59	Corporate Catering	Married
Yemi	Nigeria	Woman	40–49	Career Development Agency for Women	Married, children
Kobi	Ghana	Man	30–39	Hotel Management	Married
Kofi	Ghana	Man	40–49	Hotel Management	Married, children
Kwasi	Ghana	Woman	30–39	Chop Bar	Married, children
Maduka	Nigeria	Man	40–49	Tour Operator	Married, children
Ndidi	Nigeria	Woman	30–39	Arts Gallery Manager/ Curator	Married
Ola	Nigeria	Woman	30–39	Event Management & Catering	Married, children
Sade	Nigeria	Woman	40–49	State Council for Arts & Culture	Married, children
Shade	Ghana	Woman	20–29	Souvenir Producer and Dealer	Married
Tayo	Nigeria	Man	30–39	Tour Operator	Married
Tunde	Nigeria	Man	40–49	Tour Company	Married, children

business representatives, with the identification of subsequent participants being based on referrals from previously interviewed participants. Given recruitment took place in urban areas, we here explore the narratives of *urban-based* entrepreneurs. There are significant similarities and variations between the urban centres in Ghana and Nigeria, given their proximity to each other within West Africa and the colonial legacy in this region that ensures culture does not necessarily align with the official borders between countries. We note such similarities and variations where relevant throughout the findings and discussion. Interviews were analysed utilising thematic analysis, informed through Braun and Clarke's (2006) six-phase approach to coding and theme development. Through this process three themes were identified: 'Governance and the entrepreneurial setting', 'Negotiating hostile entrepreneurial settings' and 'The gendered dimensions of finance'. These themes form the discussion to which we now turn.

Findings

Governance and the Entrepreneurial Setting

Fostering entrepreneurship is highly dependent on a governments' ability to manage economic and social constraints, particularly for segments of society who may not have equal opportunities to participate in productive economic activities. This has led many nations to undertake significant investment in measures to counter potential barriers, such as providing grants, reducing taxes and offering low interest loans for new start-ups (Udefuna and Uzodinma 2017). Yet, whilst Ghana and Nigeria have some of the highest numbers of entrepreneurs in the world, concentrated within MSMEs in the services industries, there was a strong sense among participants of all genders that governments did not recognise this unique context as an asset:

> The government now has underplayed tourism as an industry by starting off four years ago in moving tourism as a ministry and joining it with culture and information. And of course, automatically it's been relegated to the background. It boils down from the top, that the government does not understand tourism and that is affecting every other aspect of the industry. Tourism is supposed to bring in the multiplying effect to create jobs, poverty alleviation and all that. Yes, the private sector has its own part to play, but if the government doesn't put things on the ground, it cannot work, and we cannot achieve what we want.
>
> (Ikemba, Nigeria, Man, Tourism Media, and Marketing enterprise)

Such government sentiment has materialised through taxation policies that place immense pressures on domestic start-ups whilst favouring international investment:

> We are taxed multiple times. We pay LASEPA [Lagos State Environmental Protection Agency], and we write a report for them every two years. Then we pay VAT [value added tax] to the federal government. Then we pay consumption tax, which is like the VAT.
>
> (Abeo, Nigeria, Man, Hotel enterprise)

> I know that there is a five-year tax holiday for expatriate companies. It does not make any sense that I who is a citizen, struggle to raise my own capital

> and from day one government agencies are on you to come and file your taxes. For some months, the total revenue generated was zero. So, we filed zero returns at the end of the month, but we still pay tax. So, you will realize that you cannot sustain your businesses, but the expatriate can.
>
> (Adom, Ghana, Man, Hospitality enterprise)

Strong sentiment prevailed across participant narratives that both the Ghanaian and Nigerian governments ought to undertake a more significant role in facilitating settings that would foster growth for domestic entrepreneurs:

> Startups need some tax reliefs so that when you have the business, it's like a raw idea, you need to run the business for a while before you will see all the pros and cons in it and while you are concentrating on that, then you have to be battling with tax related matters and other fixed costs. There should also be financial support for businesses, for people who have genuine business plans and when the money is given to them, their businesses should be monitored to make sure that they are in good standing i.e., they are doing the right thing and can stand on their own.
>
> (Dede, Ghana, Man, Tourism Media enterprise)

Entrepreneurial barriers rendered through governance and policy were not only economic. Nigeria, for example, is unable to meet its electricity demand, in part, due to well-documented policy failures – leading to significant impacts on the economy (Olówósejéjé 2020). A lack of reliable power supply has led to a context whereby nearly half of the country's electricity is produced using self-generated petrol and diesel generators, ensuring a significant percentage of revenue generated by MSMEs is diverted to generator running costs:

> Electricity is a big challenge. At the end of the day, we ended up having our own step-down transformer because of the electricity. For this business to stand, you need to do a lot of things because, in the country that we are in, there is no 24 hours electricity. You will need to run the generator to power your business and that will cost you money, buying diesel daily.
>
> (Abeo, Nigeria, Man, Hotel Management enterprise)

> The number one challenge is electricity. Electricity in Nigeria is so bad, we spend almost all our profit getting fuel so we can get electricity. That is one major challenge. You can imagine spending up to N4000 [$9.64US] just to fuel your generator for a day. So, you can calculate that for a month, that has taken almost all the profit in the business.
>
> (Adaku, Nigeria, Woman, Food and Beverage enterprise)

Given the importance of customer comfort and care within the services sector, ensuring a reliable power supply was a necessity for many participants. In consequence, the significant sums diverted to energy production impacted opportunities for entrepreneurs to make a profit or invest in other aspects of the business:

> We are still facing challenges of power supply. We spend an average of one million [Nigerian Naira, $2,409US] on diesel every month. …regarding the power supply, there is nothing I can do than to endure but when I looked at my profit, and I looked at my staff salaries, and everything I was forced to trim down almost by half including my staff because there was no way I could cope. That has forced some to go looking for jobs. These are people that would have been enjoying working with me if I was able to pay their salaries, they could remain with me.
>
> (Isoken, Nigeria, Woman, Corporate Catering enterprise)

Negotiating Hostile Entrepreneurial Settings

The measures implemented by both the Ghanaian and Nigerian governments was discussed by participants in relation to accessing formal finance. Both Ghana and Nigeria have high interest rates due to complex issues, including heightened inflation resulting from low manufacturing bases and low productivity (The Economist 2020). With such high interest rates, many entrepreneurs avoided accessing formal finance through banks:

> I fear loans. The interest on the loans, I fear it.
>
> (Shade, Ghana, Woman, Souvenir Producer and Dealer enterprise)

> I've never raised money from a bank. We can set out our rates where we believe you know, is commensurate to the quality and service and the product that we have. We wouldn't be able to do that if we raise money from bank, especially a Nigerian bank, because with 25 percent interest rates or more it's very crippling. You start making very desperate decisions and then, it really affects the business success.
>
> (Aisha, Nigeria, Woman, Hotel Management enterprise)

The particularities of the services industry, whereby there is often a lack of collateral security, i.e., tangible assets to offset a loan (Kimbu and Ngoasong 2013), heightened concern for some participants:

> Of course, now, in Nigeria, one of my policies as a businessperson is that I don't borrow. I have never gone to anybody. I don't believe in it because they put you under pressure you will not be prepared for, especially in our own kind of industry because of the intangibility of what you are doing.
>
> (Ikemba, Nigeria, Man, Tourism Media, and Marketing enterprise)

The fear and pressure felt by entrepreneurs regarding obtaining a loan is evidenced in the entrepreneurship literature – whereby inability to repay loans in Africa can often lead to extreme outcomes such as homelessness and imprisonment (Bateman 2010). In the absence of suitable formal finance and government assistance, nearly all participants sought to raise informal finance individually through their networks of family and friends as well as through informal microfinance institutions (cf. Ngoasong and Kimbu 2016), to establish and/or grow their enterprises:

> Working with USAID for two years, I received quite a substantial amount of money as salary. So that was something like my bedrock or something that gave me the confidence. So, I resigned from the public sector and started on my own as a tour operator.
>
> (Tunde, Ghana, Man, Tour Company enterprise)

Utilisation of personal networks was also an essential aspect in establishing and maintaining enterprise sustainability, in the absence of policy support and formal finance:

> When I was working at the guesthouse, I was making some small savings. So that was what I used. But for the drinking spot, I one day approached one of the drink distributors in our community and pleaded with him to give me the drinks to sell on credit and pay him back. I initially started with half a crate of Guinness and beer and as I sold it and returned the money to the man, I was collecting the drinks from. Gradually, I increased the number of drinks I took, and sometimes, I would sell them all by the close of the day. We had a fridge in the house, I moved it to the shop for the selling of the drinks. So, I did not take any money from anybody to start the drinking spot.
>
> (Kwasi, Ghana, Woman, Chop Bar enterprise)

> In fact, I backslid. Let me put it that way as an entrepreneur. But I have a friend who supported me financially. He injected some money into the business and helped me with marketing until things came up [improved] for me…so I had support.
>
> (Tunde, Ghana, Man, Tour Company enterprise)

The omission of formal finance opportunities, coupled with a lack of government support and immense taxation pressure, led to a dominant narrative among participants that business growth tended to be slow and organic – where change occurred incrementally, over several years:

> I did not have enough capital. So that was why I had to pause the dream again, go back into industry to work, raise capital, establish networks, and acquire an office.
>
> (Adom, Ghana, Man, Hospitality Management enterprise)

> For one and a half years I could not get started. The government wanted to kill me. LASEMA [Lagos State Emergency Management Agency] came.

> LASEPA came. Tourism whatever came. FRSC [Federal Road Safety Corps] came Lagos state remittance came even without seeing you make money. So many of them to mention. And they kept on threatening that they would lock the shop. I had not even yet started. And I just thought, which kind of government is this? And they don't even want to know the problem you are facing to help. It has been difficult.
>
> (Isoken, Nigeria, Woman, Corporate Catering enterprise)

The concept of linear entrepreneurial growth is one constructed through the dominance of Global North experiences within the entrepreneurship literature, whereby raising capital is something that is considered to predominately occur in the pre-launch phase, whilst enterprises evolve from nascent to established over time (Monteith and Camfield 2019). In this research, entrepreneurship was rather an iterative process through which enterprises were developed over several years as entrepreneurs gained access to personal savings and social networks to inject capital into the business in the absence of formal finance. Enterprises expanded, as well as contracted, as entrepreneurs negotiated other forms of work and familial and social relations over time.

More broadly, participants felt meaningful economic growth cannot be achieved without government interventions – such as tax breaks for new start-ups, low business taxes, grants and training – all of which were thought to offer enabling environments for (domestic) entrepreneurs. Without such governance infrastructure, participants felt that both the Ghanaian and Nigerian governments were somewhat hostile towards their work, in contrast to the contribution's participants felt their enterprises were making to the economy and society.

Governance and policy support, or lack thereof, were challenges experienced by entrepreneurs of all genders, rendering enduring challenges for enterprises and negatively impacting firm performance (Marlow and Patton 2005). This is a critical problem for both the Ghanaian and Nigerian governments because undercapitalisation in the start-up and development phase of a business can have enduring effects in business performance for longer term. Importantly, however, intersecting with these broader structural barriers felt by participants of all genders was the additional presence of complex and multifaceted constraints arising from gendered characterisations that produced additional barriers for women entrepreneurs. A discussion to which we now turn.

The Gendered Dimensions of Finance

Avoidance in accessing formal finance through banks was a practice carried out by many participants, across all genders. For women, however, concerns regarding high interest rates were coupled with culturally informed normative perceptions that further inhibited abilities to access funding in this way:

> Banks are, to get a loan for a woman, is difficult. Some businesses are coming up now that give soft loans, small loans to women, to small business owners, most of

> whom are women anyway. But generally, banks are usually weary in supporting a woman. There is a lot of victimization and discrimination for women.
>
> (Buki, Nigeria, Woman, Youth Development enterprise)

Here it appears that women are encountering a multiplicity of structural disadvantages in Ghana and Nigeria that present barriers to sustaining, and/or growing, their enterprises. Whilst the development of small loans has been heralded as a potentially promising avenue for providing small businesses, particularly within the services sector, with the opportunity to access formal finance (Sackeyfio 2021) – the fact that women's enterprises are concentrated in small-scale services enterprises in Ghana and Nigeria means that a two-tiered system is constructed that places women in a subordinate position whereby they are culturally hindered from applying for loans other than small loans. Moreover, adding to these gendered disparities, it has been noted that it becomes challenging to repay small loans within the context of small businesses in Africa due to the high interest rates that can cause repayments to be higher than the profit that can be returned within the enterprise (Sackeyfio 2021).

This situation ultimately reproduces gendered perceptions regarding what it means to be a woman in entrepreneurship because these structural constraints reproduce essentialist understandings that women are less effective in accessing and managing formal finance – despite these challenges being structurally and culturally produced. The complexity of these intersecting factors further hinders opportunities to trouble gendered outcomes that place women entrepreneurs in a particular cycle of disadvantage (Marlow and Patton 2005) and run counter to the aim of increasing women's decision-making and financial independence (Sackeyfio 2021). The outcome of these gendered risks is that women participants chose not to access formal finance, be it small loans or otherwise, preferring rather to finance through personal savings or to a lesser extent personal networks (Kimbu and Ngoasong 2016). Less common were examples of participants attempting to bring to question gendered constructions that hinder the ability for women to access formal finance:

> You know, I'm very focused and I'm very academic, and I was just determined. And I just felt like, you can argue all you want, but I'm going to get the funding you know, in my little pink dress.
>
> (Aisha, Nigeria, Woman, Hotel Management enterprise)

At times, challenges to such gendered hierarchies were met with backlash as women's participation in the economic sphere was felt as a disruption, or even threat, to culturally constructed roles:

> Last year, I was in a course with a particular friend, she also wants to have her own organization…but she lost her marriage to that because it became an issue of choice between this course and the husband. She was like, I will go for my course. So that was it, that was how that marriage went. So, those are some of the cultural implications that we talk about.
>
> (Buki, Nigeria, Woman, Youth Development enterprise)

Posing further challenges were gendered constructions of women as social and familial carers and providers, whereby it was understood that income ought to be prioritised to provide financial support for the extended family, as well as broad social networks, rather than being reinvested into the enterprise or saved for liquidity and emergency purposes:

> My family alone they need me more than anything… my external family is there…every month I must send money here and there… people will call Aunty; my school uniform is torn. Aunty, I have been sent home from school because of school fees and the little you have [you send through] … mobile money… ghc fifty [$6US], ghc one hundred [$12US], the highest will be ghc two hundred [$25.50US]. If I do not have it, I tell them this month is not good for me; let us see what will happen in the following month. If I get something, I share with them and I cannot stop helping them because they are my family.
>
> (Shade, Ghana, Woman, Souvenir Producer and Dealer enterprise)

> This business is all that I have. It is all I depend on financially to support myself and my family so I will not stop. I have paid my children's school fees from this business…I will not stop it… and it is good to help those who are in need…so I plan that this year I will increase the amount and add some other things like used clothing to help those in need.
>
> (Kwasi, Ghana, Woman, Chop Bar enterprise)

Considerable focus on entrepreneurship literature to the gendered expectations placed on men to perform the role of provider within Africa (cf. Ferguson 2015; Honwana 2013). Here, what is of interest is the diverse ways these expectations are increasingly placed on women as they enter the economic sphere. For women entrepreneurs in this research, expectations extended beyond the immediate family to that of the extended family and local community (cf. Kimbu and Ngoasong 2016). Managing enterprise finance, alongside upholding these gendered expectations, influenced enterprise objectives in ways that further restricted growth, particularly in the post-launch and expansion phases. The requirement to negotiate such complex gendered norms that intimately connect women's enterprises with the space of the home and run counter to assumptions informed through the Global North biased entrepreneurship literature, which posits familial support and extended social networks enhance the financial stability of enterprises (Monteith and Camfield 2019).

Conclusion

This chapter has undertaken a heterogenous approach to the study of gender, entrepreneurship and finance that seeks to examine how entrepreneurs of all genders negotiate finance. Ghana and Nigeria were taken as case studies because they offer opportunity to bring to question normative constructions of gender and finance that prevail within the entrepreneurship literature because of a prioritisation with Global North experiences. Through this approach, this chapter has brought attention to how perceptions and motivations regarding finance are informed through a complex interplay of gender, culture and governance.

The heterogenous approach to gender highlighted overlaps and variations between genders – as well as provided insights into the broader challenges influencing perceptions and motivations regarding finance within a place-specific context. More specifically it was identified how participants of all genders experienced immense challenges in negotiating formal finance within Ghana and Nigeria, which influenced decisions to avoid such finance. Such challenges were significantly influenced by a lack of government and policy support; in both Ghana and Nigeria interest rates, taxes and infrastructural costs are high, inhibiting enterprise sustainability and growth. In part, these challenges were particular to the services sector given the intangibility of service, alongside the requirement to ensure customer comfort. As a result, for Ghanaian and Nigerian enterprises in the services sector, growth tended to be iterative and non-linear, running counter to the dominant Global North narrative of linearity.

Gender was found to influence access to certain forms of formal finance, whereby women tended to be guided towards micro or small loans because of the gendered perceptions associated with women entrepreneurs and their enterprises. Such perceptions in effect produce a two-tiered system of formal finance that reinforces gendered divisions and places women in a position where it is challenging to counter gendered expectations relating to enterprise performance. It was also identified that there are expectations relating to familial and social care that influence how women entrepreneurs spend profits, potentially further inhibiting enterprise sustainability and growth.

Taken together, the above findings highlight the crucial role of government and policy in fostering entrepreneurial cultures. There is policy work to be undertaken in Ghana and Nigeria to ensure finance and taxation structures nurture entrepreneurial motivations, rather than inhibit. Part of this work is in shifting culturally informed gendered perceptions that are embedded within institutions, in very specific place-based ways. As part of shifting cultural gendered constructions, there is value in designing forms of finance that foster opportunity for gender parity through opportunity for sustainability and/or growth, rather than constructing programmes that reinstill normative constructions of women's enterprises as small scale and growth inhibited.

References

Abubakar, H. (2015). Entrepreneurship development and financial literacy in Africa. *World Journal of Entrepreneurship, Management and Sustainable Development*, *11*(4), 281–294.

Ahl, H. (2004). *The Scientific Reproduction of Gender Inequality: A Discourse Analysis of Research on Texts on Women's Entrepreneurship*. Liber AB: Sweden.

Bateman, M. (2010). *Why Doesn't Microfinance Work? The Destructive Rise of Local Neoliberalism*. London: Zed Books.

Bill and Melinda Gates Foundation. (2020). *The Full Participation Project: No Ceilings.* Accessed: 03/06/2022 from http://www.noceilings.org/entrepreneurs/.
Braun, V., & Clarke, V. (2006). Using thematic analysis in psychology. *Qualitative Research in Psychology, 3*, 77–101.
Doran, J., McCarthy, N., & O'Connor, M. (2018). The role of entrepreneurship in stimulating economic growth in developed and developing countries. *Cogent Economics and Finance, 6*(1).
Ferguson, J. (2015). *Give a Man a Fish: Reflections on the New Politics of Distribution.* Durham, NC and London: Duke University Press.
Figueroa-Domecq, C., de Jong, A., & Williams, A. (2020). Gender, tourism, and entrepreneurship: A critical review. *Annals of Tourism Research*, 84.
Hill, F., Leitch, C., & Harrison, R. (2006). 'Desperately seeking finance?' The demand for finance by women-owned and -led businesses. *Venture Capital, 8*(2), 159–182.
Honwana, A.M. (2013). The time of youth: Work, social change, and politics in Africa. *Social Dynamics, 39*(3), 554–557.
Huysentruyt, M. (2014). Women's social entrepreneurship and innovation. In *OECD Local Economic and Employment Development (LEED) Papers*. OECD Publishing.
Kim, G. (2006). Do equally owned small businesses have equal access to credit? *Small Business Economics, 27*, 369–386.
Kimbu, A.N., & Ngoasong, M.Z. (2013). Centralized decentralization of tourism development: A network perspective. *Annals of Tourism Research, 40*, 235–259.
Kimbu, A.N., & Ngoasong, M.Z. (2016). Women as vectors of social entrepreneurship. *Annals of Tourism Research, 60*, 63–79.
Kwapisz, A., & Hechavarría, D. (2018) Women don't ask: An investigation of start-up financing and gender. *Venture Capital, 2*, 159–190.
Leitch, C., Welter, F., & Henry, C. (2018). Women entrepreneurs' financing revisited: Taking stock and looking forward. *Venture Capital, 20*(2), 103–114.
Marlow, S. (2013). Exploring future research agendas in the field of gender and entrepreneurship. *International Journal of Gender, and Entrepreneurship, 6*(2), 102–120.
Marlow, S., & Patton, D. (2005). All credit to men? Entrepreneurship, finance, and gender. *Entrepreneurship Theory and Practice, 29*(6), 717–735.
Marlow, S., & Swail, J. (2014). Gender, risk, and finance: Why can't a woman be more like a man? *Entrepreneurship & Regional Development, 26*(1–2), 80–96.
Mastercard. (2018). *Index of Women Entrepreneurs.* Accessed: 03/06/2022 from https://www.mastercard.com/news/ap/en/media-library/infographics/2018-mastercard-index-of-women-entrepreneurs/
Meunier, F., Fantoni, S., & Kouhlani-Nolla, S. (2022). *We-Data: Measuring the Gap in Female Entrepreneurship around the World.* Accessed: 03/06/2022 from https://blogs.worldbank.org/developmenttalk/we-data-measuring-gap-female-entrepreneurship-around-world.
Ministry of Trade, Ghana. (2019). *National Micro, Small and Medium Enterprises (MSME) Policy.* Accra: Government of Ghana.
Monteith, W., & Camfield, L. (2019). Business as family, family as business: female entrepreneurship in Kampala, Uganda. *Geoforum, 101*, 111–121.
Ngoasong M.Z., & Kimbu, A.N. (2016). Informal microfinance institutions and development-led tourism entrepreneurship. *Tourism Management, 52*, 430–439.
Ngoasong, M.Z., & Kimbu, A.N. (2019). Why hurry? The slow process of high growth in women-owned businesses in a resource-scarce context. *Journal of Small Business Management, 57*(1), 40–58.

Obeng, B., & Blundel, R. (2015). Evaluating enterprise policy interventions in Africa: A critical review of Ghanaian small business support services. *Journal of Small Business Management*, *53*(2), 416–435.

Olówósejéjé, A.S. (2020). What Nigeria's poor power supply really costs and how a hybrid system could work for business. *The Conversation*. Accessed: 03/06/2022 from https://theconversation.com/what-nigerias-poor-power-supply-really-costs-and-how-a-hybrid-system-could-work-for-business-144609.

PwC. (2022). *Nigeria SME Survey*. Accessed: 03/06/2022 from https://www.pwc.com/ng/en/events/nigeria-sme-survey.html.

Sackeyfio, N. (2021). Gender, microfinance, and microcredit in West Africa: Empowerment for Whom? *New Political Science*, *43*(1), 86–105.

The Economist. (2020). *Why Interest Rates are so High in Africa*. Accessed: 05/06/2022 from https://www.economist.com/finance-and-economics/2020/05/21/why-interest-rates-are-so-high-in-africa.

Udefuna, P., & Uzodinma, E. (2017). Entrepreneurship development in Nigeria: Issues for policy and legislative attention. *Specialty Journal of Politics and Law*, *2*(4), 33–44.

8 Ethnotourism and Buen Vivir

Gendered Pathways to Sustainable Tourism Development in Colombia

Andrés Morales Pachón and Sara Calvo Martinez

Introduction

As an academic endeavour, this chapter delves into the nuanced intersections of ethnotourism and the indigenous philosophy of BV, exploring how these concepts serve as pivotal elements for fostering sustainable development and preserving cultural integrity within indigenous and ethnic communities. This introduction aims to underscore the intricate relationships between tourism, community involvement, environmental stewardship, and cultural preservation, framed through the lens of Buen Vivir (BV) – a worldview that advocates for a harmonious existence among humans and with nature.

Ethnotourism, a subset of tourism that emphasises engaging with and understanding the cultural heritages of indigenous and ethnic groups, has gained prominence as a sustainable approach to tourism. It offers a profound context for visitors to experience and appreciate the rich cultural fabrics and traditions that define various communities around the globe. Unlike conventional tourism models, which often risk undermining local cultural identities and ecological balances, ethnotourism presents a pathway that potentially enhances socio-economic development while ensuring the conservation of both cultural and natural heritage. This chapter critically evaluates ethnotourism capacity to align with the sustainable tourism criteria – balancing economic viability, environmental protection, and social equity.

The concept of BV, or "Good Living," rooted in the knowledge systems of Latin America's indigenous peoples, particularly those in the Andean regions, proposes a radical shift from traditional development paradigms that are predominantly centred on economic growth and material accumulation. Instead, BV promotes a comprehensive approach to development that values community solidarity, reciprocity, and a balanced coexistence with nature. This philosophy challenges the current global development agenda and offers an alternative that prioritises well-being, ecological sustainability, and communal prosperity over individual wealth and consumption.

Integrating BV with ethnotourism provides a unique opportunity to reconsider how tourism is practised. It allows for a model that not only supports economic development but also fosters environmental stewardship and bolsters cultural preservation. This chapter provides several illustrative examples, such as the

DOI: 10.4324/9781003315858-8

case of Asociación Dugjin in Colombia, where ethnotourism initiatives driven by indigenous philosophies have enabled communities to thrive while maintaining their cultural and environmental ethos.

One of the core discussions in this chapter revolves around the role of governments and international bodies in promoting and regulating ethnotourism. It critically examines the policies that have been implemented to support ethnotourism and scrutinises their effectiveness in aligning with the principles of BV. The analysis reveals a complex tapestry of challenges and opportunities, where policy frameworks often lag in accommodating the dynamic interplay between preserving cultural values and promoting sustainable tourism practices.

Furthermore, this chapter addresses the potential pitfalls of ethnotourism, including the risks of commodification of culture and over-tourism, which can lead to the dilution of cultural identities and environmental degradation. It proposes a set of criteria and best practices for ethnotourism that respect and uphold the principles of BV, ensuring that tourism development does not come at the expense of cultural and ecological integrity.

Theoretical contributions to the field of sustainable tourism and indigenous studies are also a focal point of this chapter. It engages with various academic discourses to contextualise ethnotourism within broader socio-economic and environmental frameworks. By weaving together theoretical insights with practical examples, this chapter offers a robust analysis of how ethnotourism can be effectively implemented to support the aspirations of indigenous and ethnic communities while contributing to the global sustainability agenda.

This chapter not only explores the synergies between ethnotourism and BV but also critically assesses their potential to redefine the landscape of global tourism and development. Through a detailed examination of case studies and policy analyses, it provides valuable insights into how adopting a BV approach can enhance the sustainability and cultural sensitivity of tourism practices, ultimately leading to more equitable and inclusive development outcome. This introduction sets the stage for a comprehensive exploration of these themes, encouraging readers to reflect on the transformative potential of integrating indigenous philosophies with contemporary tourism and development practices.

Literature Review

Defining Ethnotourism and Sustainable Tourism: Concepts and Contexts

Ethnotourism is a distinctive sector within the global tourism market that centres on the cultural heritage and traditional practices of Indigenous and ethnic communities. This form of tourism is characterised by the interaction between tourists and local communities, which facilitates deep engagement with cultural expressions and lifestyles. Mukatova et al. (2022) underscore that ethnotourism not only promotes cultural preservation but also contributes to sustainable development by encouraging local participation and generating economic benefits for regional economies. Activities typical of ethnotourism include participating in traditional ceremonies, festivals, and communal practices, allowing tourists to experience authentic cultural settings and fostering mutual respect and cultural exchange. This engagement often leads to a more profound appreciation of the host communities' cultural heritage, contributing to its sustainability.

Sustainable tourism, as outlined by Butler (1999), is the development and maintenance of tourism activities in such a way that they remain viable over an indefinite period without degrading the environment or socio-cultural fabric. This concept emerged from the broader discourse on sustainable development, particularly influenced by the Brundtland Commission's report, "Our Common Future." Sustainable tourism necessitates a balanced approach that safeguards the interests of both current and future generations. It integrates environmental conservation with socio-cultural enrichment and economic viability, ensuring that tourism practices do not compromise local cultures, traditions, or environmental integrity. A critical issue within the sustainable tourism discourse is the need for clear definitions and robust frameworks to assess and implement sustainable practices effectively. Butler (1999) highlights that without rigorous standards, the term "sustainable" can be misapplied, leading to practices that may not truly support sustainable development goals.

An illustrative example of ethnotourism from a developed country is the Maori cultural experiences offered in New Zealand. Tourists can engage in traditional Maori ceremonies, visit marae (meeting grounds), and learn about Maori history and customs. These activities provide a profound insight into Maori culture while promoting cultural preservation and economic benefits for Maori communities. In Latin America, Peru offers community-based tourism initiatives in the Andean regions, where visitors can engage with Indigenous communities, participate in traditional farming practices, and learn about Andean culture and heritage. These initiatives exemplify how ethnotourism can provide meaningful cultural exchanges, support sustainable development, and ensure the preservation of cultural heritage (Mukatova et al., 2022). Both examples highlight the potential of ethnotourism to foster cultural appreciation and economic growth while adhering to the principles of sustainability.

The Role of Indigenous Communities in Ethnotourism

Indigenous communities engage in various entrepreneurial activities related to tourism that contribute to cultural preservation, economic empowerment, and environmental stewardship. These activities include offering guided tours that showcase their traditional practices, creating and selling handicrafts, and providing authentic cultural experiences through festivals, ceremonies, and communal events. For instance, in Mexico, rural Indigenous communities within Natural Protected Areas (NPAs) have leveraged tourism as a sustainable development intervention since the 1980s (Espeso-Molinero & Pastor-Alfonso, 2020). The community of Santa Cruz Tepetotutla in Oaxaca exemplifies this approach, where traditional social organisation plays a crucial role in enhancing collective well-being, conserving biodiversity, and preserving cultural heritage through tourism initiatives.

Women play a pivotal role in the tourism-related entrepreneurial activities within Indigenous communities, significantly influencing gender dynamics and socio-economic structures. Women often lead in the creation and sale of handicrafts, culinary offerings, and hospitality services, thereby contributing to household incomes and community economies. Their involvement not only empowers them economically but also fosters social cohesion and cultural continuity. In many Indigenous societies, women are the custodians of cultural knowledge and practices, and their participation in tourism helps to transmit these traditions to visitors, reinforcing their cultural identity. This dynamic is crucial for promoting gender equity and enhancing the overall socio-economic fabric of Indigenous communities (Harbor & Hunt, 2021; Pereiro, 2019). Collaborative efforts and culturally sensitive approaches are essential to ensure that women's roles are recognised and supported, thereby enriching the tourism experience and promoting sustainable community development.

Conceptualising BV for Understanding Ethnotourism

BV (Buen Vivir), or "Good Living", is a transformative and decolonial paradigm that fundamentally challenges traditional Western concepts of development and progress. Rooted in the rich tapestry of Latin American indigenous worldviews, particularly the Andean cosmovision, BV advocates for a harmonious coexistence between humans and nature and among humans themselves. This conceptual framework is imbued with values of community, solidarity, reciprocity, and an intrinsic respect for the natural world, setting it apart from conventional development models that prioritise individualism and the exploitation of natural resources. BV's emphasis on collective well-being, ecological balance, and cultural integrity offers a holistic approach to development that is sustainable and equitable. Scholars such as Acosta (2015), Cabildo Yanacona (2009), and Dávalos (2011) have extensively explored BV, highlighting its potential to reshape our understanding of development in a way that respects and preserves the diverse cultural and ecological landscapes of Latin America. This shift challenges the dominant paradigms that have historically led to environmental degradation and social inequality, promoting instead a vision of development rooted in ethical principles and mutual respect.

Figure 8.1, which conceptualises BV, visually represents the three distinct yet interconnected approaches: indigenist, socialist/statist, and ecologist/developmentalist. Each approach contributes unique perspectives and strategies for realising the principles of BV within contemporary societies. The indigenist approach focuses on the resurgence and recognition of indigenous identities, knowledge, and governance systems. It advocates for a plurinational state structure that acknowledges the sovereignty and self-determination of indigenous peoples within a single national polity. This approach emphasises the revival of ancestral practices and wisdom as foundational to constructing a more inclusive and equitable society (Hidalgo-Capitán & Cubillo-Guevara, 2017). This revival is critical in empowering indigenous communities to reclaim their cultural heritage and influence national policies

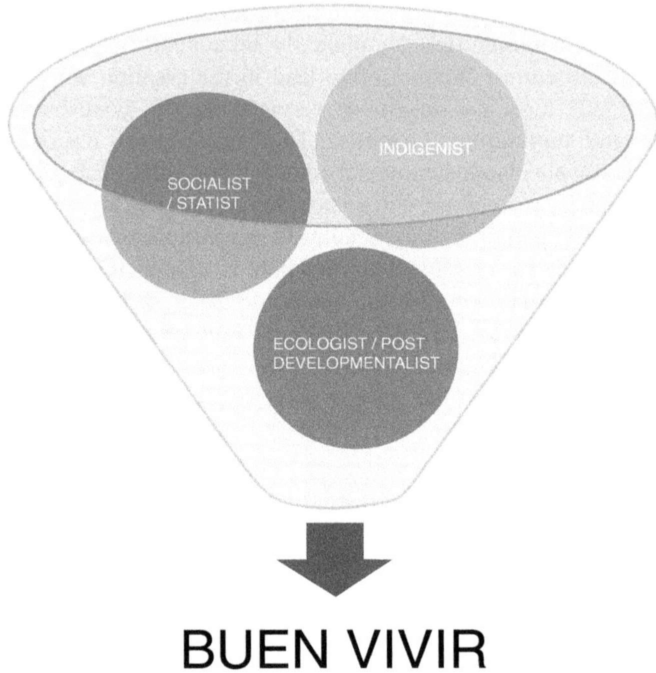

Figure 8.1 The conceptualisation of BV

in ways that promote social justice and ecological balance. The socialist/statist approach, on the other hand, aims to address social inequities through state-led initiatives inspired by neo-Marxist theories. It seeks to reorient state policies and economic models towards achieving greater social justice, often through redistributing resources and implementing reforms that incorporate elements of indigenous philosophies (Hidalgo-Capitán & Cubillo-Guevara, 2016).

The ecologist/developmentalist approach critiques conventional development paradigms from an ecological and post-development perspective. It foregrounds biocentrism and the rights of nature, advocating for a radical transformation of societal values and economic practices to ensure ecological sustainability and respect for all forms of life. This approach challenges the anthropocentric worldview that underpins many development models and calls for a shift towards practices that prioritise ecological integrity and the intrinsic value of all living beings. Together, these approaches form a comprehensive framework for BV that integrates indigenous wisdom with contemporary social, political, and ecological thought. By envisioning development as a process of achieving communal well-being, ecological balance, and cultural diversity, BV offers a compelling alternative to the dominant paradigms of development that have led to environmental crises and social fragmentation (Hidalgo-Capitán & Cubillo-Guevara, 2015). This comprehensive framework not only critiques the existing development models but also provides practical strategies for creating more sustainable and equitable societies.

Figure 8.2 introduces the BV model's values and pillars, which include community, solidarity and reciprocity, and harmony and complementarity. These values inform the six pillars: Rights of Nature, community well-being, decolonisation, plurinational state, economic pluralism, and democratisation. This comprehensive framework integrates indigenous wisdom with contemporary social, political, and ecological thought, offering an alternative to dominant development paradigms that have contributed to environmental and social crises. The Rights of Nature pillar recognises nature as a subject with intrinsic rights, challenging the conventional view of the environment as a resource to be exploited. This radical rethinking of environmental governance calls for legal and policy frameworks that protect ecosystems and biodiversity as integral members of the community. Community well-being focuses on enhancing the quality of life for all members of society, advocating for development policies that prioritise social equity, health, education, and cultural preservation. Decolonisation involves challenging colonial legacies and creating models grounded in indigenous philosophies and values, while the plurinational state pillar promotes inclusive governance. Economic pluralism supports diverse economic systems, and democratisation emphasises participatory governance (Morales, 2019).

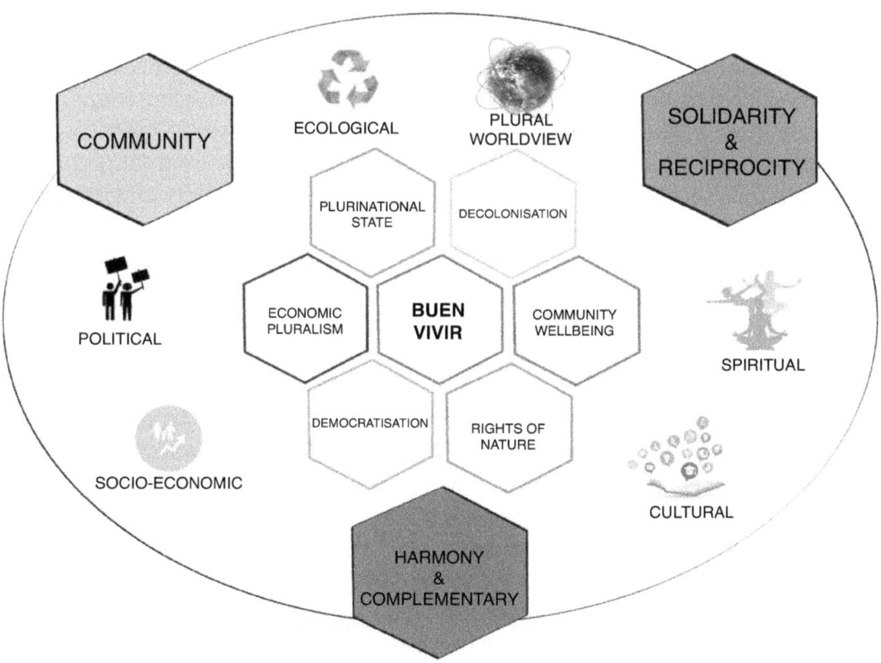

Figure 8.2 The BV model (values and pillars)

Ethnotourism aligns closely with the principles of BV as it emphasises cultural preservation, economic empowerment, and environmental stewardship. By engaging tourists in authentic cultural experiences, ethnotourism fosters mutual respect and cultural exchange, which are essential for the sustainability of indigenous communities. For example, community-based tourism initiatives in Peru's Andean regions enable visitors to participate in traditional farming practices and learn about local culture, promoting economic benefits while preserving cultural heritage. These initiatives illustrate how tourism can be a tool for sustainable development that aligns with the values of BV. Such tourism projects can strengthen local economies by creating jobs and generating income and also promoting the conservation of cultural traditions and natural environments (Espeso-Molinero & Pastor-Alfonso, 2020). This dual focus on economic empowerment and cultural preservation helps to ensure that tourism development benefits local communities without compromising their cultural and ecological integrity.

Women in indigenous communities play a crucial role in tourism-related entrepreneurial activities, contributing significantly to household incomes and community economies. They often lead in creating and selling handicrafts, culinary offerings, and providing hospitality services. Their involvement not only empowers them economically but also reinforces social cohesion and cultural continuity. Supporting women's roles in tourism can enhance gender equity and strengthen the socio-economic fabric of indigenous communities, aligning with the broader goals of BV and sustainable tourism. Harbor and Hunt (2021) and Pereiro (2019) highlight the importance of recognising and supporting women's contributions to tourism, which can lead to more inclusive and sustainable development outcomes. By acknowledging the significant impact women have on tourism, communities can create more balanced and equitable economic opportunities. This recognition and support can help to ensure that women are fully integrated into the economic benefits of tourism development, thereby promoting gender equity and enhancing the overall well-being of the community.

In essence, BV represents a paradigm shift in our understanding of development, offering a holistic and integrative framework that prioritises the well-being of all members of the community, including the natural environment. By grounding development in indigenous values and principles, BV provides a compelling alternative to the unsustainable and inequitable models that have dominated global development discourse. Ethnotourism, as an application of BV principles, demonstrates how tourism can be a tool for sustainable development that respects and preserves cultural heritage, promotes economic empowerment, and fosters environmental stewardship. Supporting the roles of women in tourism further enhances these outcomes, contributing to more inclusive and equitable development. By embracing the values and pillars of BV, communities can create a more just, sustainable, and harmonious world. This holistic approach to development not only addresses the immediate needs of communities but also ensures the long-term sustainability and resilience of both human and natural systems.

Method: Participatory Video Research for Understating Indigenous Practices

This study employs a participatory video research (PVR) methodology, rooted in Morales' (2019) work and ongoing research in Latin American countries. The research design is a multiple case study approach, focusing on five Indigenous Community-Based Organisations (ICOs) in Colombia: Misak, Yanacona, Curripaco, Puinave, and Wayuu. PVR was selected to empower participants and involve them directly in the research process, allowing for the collection of rich, multimodal data reflective of their lived experiences. Data collection methods included focus groups, interviews, and participatory video-making, which facilitated a comprehensive understanding of each ICO's unique context and challenges (Blazek & Hraňová, 2012). This approach aligns with the principles of participatory action research, ensuring that the research is grounded in the perspectives and priorities of the indigenous communities involved. The methodology emphasises the importance of community involvement and the value of indigenous knowledge systems in shaping sustainable development practices (Tremblay & Peredo, 2014).

The participatory video aspect of the research not only empowers community members but also enriches the data collected, providing a visual and narrative depth that traditional methods might miss (Mistry & Berardi, 2012). By integrating PVR with focus groups and semi-structured interviews, the study captures the everyday realities of the ICOs, offering a holistic understanding of their activities and impact. The insights gained from this research are particularly relevant to the field of ethnotourism, highlighting the potential for tourism to promote cultural preservation and sustainable development (Espeso-Molinero & Pastor-Alfonso, 2020). By ensuring that the narratives and experiences of indigenous communities are central to the research, this approach aligns with the principles of BV, promoting a model of development that prioritises ecological balance, cultural integrity, and collective well-being (Morales, 2019). The findings underscore the importance of community participation in tourism initiatives, demonstrating how ethnotourism can be leveraged to support the economic and social needs of indigenous communities while preserving their cultural heritage (Harbor & Hunt, 2021; Pereiro, 2019).

A Case Study of Asociación Dugji

Asociación Dugjin, established in 2016, is a community-led initiative within the Puinave community of Inírida Guainía, Colombia. Named after a concept symbolising the creation and origin of Puinave culture, Asociación Dugjin aims to leverage ethnotourism for community development and cultural preservation. Ethnotourism is an alternative to mainstream tourism, which is typically dominated by foreign corporations, and seeks to provide economic opportunities while promoting the conservation and appreciation of the community's cultural heritage and natural resources. This is achieved through tourism that celebrates cultural wealth and immerses visitors in the local landscapes and attractions.

The association consists of 20 member families who actively participate in its operations and decision-making processes. The stakeholders extend beyond these immediate members to include the wider Puinave community, tourists, and clients. The association's impact includes generating employment opportunities, promoting Puinave culture, and fostering community empowerment. They work closely with local authorities to ensure that all activities align with cultural and environmental sustainability, seeking permissions, and integrating community feedback into their planning and operations. One member highlighted the significance of their approach: "We make decisions unanimously.... That's why it is so important to make mingas of thinking" [Member].

Governance Structure

The governance of Asociación Dugjin is unique, characterised by an unanimity-based decision-making process involving 20 member families. This method ensures that decisions reflect the collective will of the community, fostering an inclusive and participatory environment. General assemblies are pivotal, providing a platform for extensive discussions and collective decision-making. Decisions are made only when there are no objections, which can sometimes involve lengthy deliberations to ensure consensus. As stated by the president, "Our decisions ar discussed by everyone in our communities in our morning assenblies, when I mean everyone is everyone. Even children, our territory is sacred, and we all have a say, the community's perspective is so important."

The integration of Dugjin's operations within their traditional territories underscores their commitment to cultural preservation and sustainable development.

> All our services are designed to show the tourist about our culture our territory is beautiful and there are lots of place to visit.... before we started our organisation, we asked our authorities for permission to set up a business... we explain them that it was important for the community, and we will have many advantages for all in the long term .... at the end all the tourist services offered are in our resguardo thus we need to inform the whole community and ask for permission to our authorities.
>
> [Vice-President]

The above governance structure helps maintain a unified and cooperative organisational culture. To ensure the effective operationalisation of governance, there are three key committees and leadership roles:

General Assembly. The assembly is the primary decision-making body where all members participate in discussions and decisions. The assembly is led by xx and ... A community member stated,

> All of our decisions have to go through assembly, our territory is so important and sacred, all of our community have to agree who visits and what activity we are going to do, even children have a say in this process.

The assembly is the primary decision-making body where all members participate in discussions and decisions. The Assembly is led by the older members of the community, and decisions are made through a process that values consensus and inclusivity. The organisation's assemblies are conducted as needed (e.g., lack of income, business strategies, or board appointments), and when there is disagreement, a democratic vote is conducted to ensure that the majority view prevails. In the event of a tie or unresolved conflict, the eldest member has the authority to make the final decision, ensuring that the assembly can move forward effectively. However, when the whole community is involved in the assemblies (as it might create an indirect or direct impact on the territory), decisions must be unanimous. The elders of the organisation must ensure that the whole community agrees with any decision that might affect the territory (e.g., a new tour, new touristic activity, and increased number of tourists).

A community member stated,

> All of our decisions have to go through assembly, our territory is so important and sacred, all of our community have to agree who visits and what activity we are going to do, even children have a say in this process.

The reference to "our territory" in this quotation relates to its deep cultural and spiritual significance. The territory is considered sacred as it embodies the heritage, history, and spiritual beliefs of the community. This sacredness influences their stringent decision-making process regarding visitors and activities, which is also reflected in the role of tour guides who educate visitors about the cultural and historical importance of the land.

Executive Committee. Comprised of elected leaders who oversee daily operations and strategic initiatives. A typical strategic initiative that is relevant to tourism is the development of eco-tourism trails. This initiative was proposed to promote sustainable tourism while preserving the natural environment and cultural heritage. The proposal was presented to the assembly, where it underwent thorough discussion involving all community members. Concerns about environmental impact and cultural sensitivity were raised and addressed. After extensive deliberation and a democratic vote, the assembly decided to proceed with the eco-tourism trails, ensuring that guidelines were established for minimal environmental disruption and cultural respect. This decision highlights the leadership role of the Executive Committee in balancing tourism development with community values and sustainability goals.

Operational Committees. Focus on specific areas such as guided tours, accommodation, and transport. An example of a specific area that was discussed during the field is the operation of guided tours for international tourists. One such tour was organised to visit the sacred waterfall site, a place of great cultural and spiritual significance. The tour provided an immersive experience into the community's heritage and environmental conservation efforts. Each tourist was charged around US$150, which covered the guide fee, food, transportation, and a contribution to the community's preservation fund. The operational committee meticulously planned the tour, coordinating logistics, ensuring the availability of knowledgeable guides, and preparing educational materials. They held several meetings to finalise the itinerary, address potential concerns, and ensure that the tour adhered to the community's values and sustainability principles.

The association also integrates traditional knowledge and practices into its governance, reflecting the Puinave cosmovision known as muriutún, which emphasises harmony with nature and principles of reciprocity and solidarity. As stated by one member, "All our services are designed to show the tourist about our culture, our territory is beautiful and there are lots of places to visit."

Gender Dynamics

Gender dynamics within Asociación Dugjin are particularly noteworthy, reflecting broader socio-economic conditions and cultural practices within the Puinave community. Women in this community often face socio-economic exclusion, which limits their access to economic opportunities and participation in decision-making processes. However, ethnotourism has provided a unique platform for empowerment by offering avenues to improve their income and actively engage in community governance.

Women play a crucial role in the operations of Asociación Dugjin, contributing significantly to various activities such as traditional medicine, local gastronomy, and hospitality services. This involvement not only enhances their economic status but also reinforces their importance in preserving and transmitting cultural heritage. For example, the Operational Committee, which focuses on guided tours, accommodation, and transport, is led by a woman who has been instrumental in organising eco-tourism initiatives. Additionally, one of the successful guided tours to the sacred waterfall site was led by a woman, showcasing her deep knowledge and leadership in cultural and environmental education. These examples highlight the vital contributions and leadership roles women hold within the association.

In the decision-making processes of Asociación Dugjin, women are equally important. The unanimity-based decision-making model ensures that all voices, including those of women, are heard and respected. For instance, a suggestion by a woman to create a traditional medicine garden was adopted by the association, highlighting her expertise in this domain and reinforcing the community's commitment to preserving cultural practices. This inclusive approach empowers women by recognising their contributions and ensuring their perspectives are integral to the association's operations and strategies. One member emphasised this participatory nature, stating, "We make decisions unanimously … That's why it is so important to make mingas of thinking" [Vice-President].

By participating in the association, women gain leadership experience and influence over community affairs, challenging traditional gender roles and promoting gender equity. In situations where health and wellness initiatives require gender role recognition, women are often expected to take the lead. For example, they organise and manage health education workshops and traditional medicine practices, ensuring the well-being of the community (even though the role of shaman, or indigenous doctor, has historically been performed by men). The involvement of women in the decision-making process is not only a reflection of their importance in indigenous communities but also a critical factor in the success of Asociación Dugjin. Their contributions help ensure that the association's initiatives are holistic, culturally appropriate, and sustainable.

Environmental and Cultural Integration

Dugjin's operations are deeply connected to their traditional territories, emphasising the preservation of cultural heritage and sustainable development. Their services, which include guided tours, accommodation, camping areas, traditional medicine, local gastronomy, and river transportation, are designed to showcase and respect their culture and environment. The association integrates traditional ecological knowledge with some Western techniques, maintaining a critical perspective to ensure sustainability aligns with their cultural values. This is highlighted by their commitment: "We have to acknowledge that there is already an embedded Western ideology in our practices… what I want to say is that there is a very fine line between organisational sustainability and the environmental one" [Treasurer]. One specific Western ideology that the community has adapted is the concept of sustainable tourism, which emphasises the importance of minimising environmental impact while promoting local culture and economy (Harbor & Hunt, 2020).

Funding and Support

The establishment and growth of Asociación Dugjin have been supported by external institutions, which provided essential resources for setting up the association (e.g., Cotregua, a cooperative bank with microcredit, and grants from the Organización Nacional Indígena de Colombia (ONIC) and Organizacion Nacional de los Pueblos Indigenas de Colombia (OPIAC)). Although not always financial, this support covered various costs and helped formalise the association, enabling it to access additional resources and expand its initiatives. Formalisation involved registering the association as a community non-profit organisation with the relevant government agency, which in this case is the local regional authority. This process granted the association legal recognition and the ability to apply for grants and support from various institutions. A member noted, "By formalising ourselves we could get access to external help … there were few institutions that contributed to the establishment of our association." [President].

Cultural and Socio-economic Development

The impact of Asociación Dugjin reaches beyond its immediate members to the wider Puinave community, tourists, and clients. Despite being in its early stages, the association has already started generating employment and promoting Puinave culture, thus meeting its broader objectives of economic support and cultural celebration. One example of an activity undertaken by Asociación Dugjin to promote Puinave culture is the annual Puinave Cultural Festival, which showcases traditional dances, music, and crafts. This festival attracts both domestic and international tourists, providing a platform for cultural exchange and appreciation. These cultural activities, apart from having spiritual effects with regard to harmony within the communities, are illustrative of outreach events that attract tourists (Ngoasong & Kimbu, 2016).

The able initiative also exemplifies a holistic approach to ethnotourism where economic development, cultural preservation, and community empowerment are interlinked. A member also emphasised the employment creation component of community impact, stating, "Why… why do we organise? in order to create employment for ourselves and the community…. Because there were not employment opportunities here in the resguardo…." [Member].

Conclusion

This chapter critically assesses the integration of BV with ethnotourism, highlighting its transformative potential for sustainable and equitable development in indigenous and ethnic communities. Through the lens of the Asociación Dugjin case study and other practical examples, the discussion underscores how ethnotourism can be a conduit for promoting cultural preservation, community empowerment, and environmental stewardship. Importantly, this chapter identifies BV not just as a philosophical ideal but as a viable alternative to mainstream tourism practices, which often overlook the deeper cultural and ecological implications of development.

The activities of community groups like Asociación Dugjin significantly promote sustainable tourism by embedding local cultural values and environmental stewardship into their operations. Through initiatives such as guided tours, traditional medicine, and local gastronomy, they ensure that tourism activities are closely aligned with the preservation of cultural heritage and natural resources. This approach not only enriches the visitor experience but also supports the sustainable use of resources, ensuring that tourism development does not come at the expense of environmental integrity.

Gender dynamics play a crucial role in the equitable development of these communities. In Asociación Dugjin, women are empowered to participate actively in both the operational and decision-making processes. This inclusion helps address socio-economic exclusion and enhances their economic status, allowing them to contribute significantly to community development. Women's involvement in key areas such as traditional medicine, hospitality, and cultural education underscores their importance in preserving and transmitting cultural heritage, fostering a more inclusive and equitable community structure.

Social policies can encourage greater participation of women at different levels within community groups. Policies aimed at providing education, vocational training, and leadership development for women can further enhance their involvement and impact. In communities where decision-making is already led by women, such as Asociación Dugjin, it is important to celebrate these leadership roles. These examples offer valuable lessons on the benefits of inclusive governance, demonstrating that when women are integral to decision-making processes, the outcomes tend to be more holistic and sustainable. Recommendations include fostering environments that support women's leadership, ensuring equal access to resources, and promoting gender-sensitive policies that encourage active female participation in all aspects of community and economic life.

Through such scholarly contributions, we pave the way for future research that can further explore and refine the integration of indigenous philosophies like BV with global tourism practices, ensuring that they contribute meaningfully to the well-being of all stakeholders involved. This synthesis of theory and practice not only enriches our understanding of tentorium's role in sustainable development but also advocates for more culturally sensitive and environmentally sustainable tourism practices globally.

References

Acosta, A. (2015). El Buen Vivir como alternativa al desarrollo. *Rev. Política Soc.*, *52*(2), 299–330.

Blazek, M., & Hraňová, P. (2012). Emerging relationships and diverse motivations and benefits in participatory video with young people. *Children's Geographies*, *10*(2), 151–168. https://doi.org/10.1080/14733285.2012.667917

Butler, R. W. (1999). Sustainable tourism: A state-of-the-art review. *Tourism*, *1*, 7–25.

Cabildo Yanacona. (2009). *Sumak Kawsay Kapak Ñan: Plan de Salva Guarda del Pueblo Yanacona*. Cabildo Yanacona, Popayan.

Dávalos, P. (2011). Sumak Kawsay (La vida en plenitud). *Presented at the Convivir para perdurar: Conflictos ecosociales y sabidurías ecológicas*, Icaria, pp. 201–214.

Espeso-Molinero, P., & Pastor-Alfonso, M. J. (2020). Governance, community resilience, and indigenous tourism in Nahá, Mexico. *Sustainability*, *12*, 5973.

Harbor, L. C., & Hunt, C. A. (2020). Indigenous tourism and cultural justice in a Tz'utujil Maya community, Guatemala. *Journal of Sustainable Tourism*, *29*(2–3), 214–233.

Hidalgo-Capitán, A., & Cubillo-Guevara, A. (2015). La Trinidad Del Buen Vivir En Ecuador. *Política Exter. Latinoam. Análisis*, 14.

Hidalgo-Capitán, A. L., & Cubillo-Guevara, A. P. (2016). *Transmodernidad y transdesarrollo. El decrecimiento y el buen vivir como dos versiones análogas de un transdesarrollo transmoderno*. Ediciones Bonanza, Huelva, España.

Hidalgo-Capitán, A. L., & Cubillo-Guevara, A. P. (2017). Deconstrucción y genealogía del buen vivir latinoamericano. El (trino) buen vivir y sus diversos manantiales intelectuales. *Int. Dev. Policy Rev. Int. Polit. Dév.*, 9.

Mistry, J., & Berardi, A. (2012). Bridging indigenous and scientific knowledge Local ecological knowledge must be placed at the center of environmental governance. *Science*, 352(6291), 1274–1275. DOI: 10.1126/science.aaf1160

Morales, A. (2019). *Indigenous-Hybrid Organisations in Colombia: A Multi-Level Analysis within the Buen Vivir Model*. Open University, UK, Milton Keynes.

Mukatova, R., Mussina, K., & Rodríguez, M. (2022). Scientific approaches to the definition of ethno-tourism concept. *Eurasian J. Econ. Bus. Stud.*, *65*, 47–59.

Ngoasong, M.Z., & Kimbu, A.N. (2016). Informal microfinance institutions and development-led tourism entrepreneurship. *Tourism Management, 52*, 43w0–439.
Pereiro, X. (2019). A review of Indigenous tourism in Latin America: Reflections on an anthropological study of Guna tourism (Panama). *Sustain. Tour. Indig. Peoples*, 55–72.
Tremblay, C. & Peredo, A. (2014). Participatory video as an approach for strengthening collective social entrepreneurship: The recycling cooperative movement in Brazil. In *Social Entrepreneurship and Research Methods (Research Methodology in Strategy and Management, Vol. 9)*, 189–214. https://doi.org/10.1108/S1479-838720140000009015

9 Towards a Framework for Enabling Resilience for Sustainability in Women-Owned Tourism Businesses

Michael Zisuh Ngoasong

Introduction

Whereas scholars have researched how women entrepreneurs build their resilience (Matharu & Juneja, 2021; Welter et al., 2018), this chapter proposes an alternative view of resilience as a capacity of women entrepreneurs to achieve sustainability for their business. Traditionally, this construct has focused on bouncing back to the old normal. However, this traditional view of resilience does not seem relevant in the conditions of liminality – the conditions where old systems (e.g., rules, norms, regulations, and procedures) are ineffective, whereas the new ones are not yet institutionalized post-crisis. To this end, this chapter suggests an alternative view of resilience that addresses the dilemma women-owned businesses face while coping with the pandemic: either to bounce back or to bounce forward. Thus, instead of aiming to return to the old ways of working (bouncing back), women entrepreneurs may opt for transforming (bouncing forward). This chapter argues that sustainability (future value) and resilience (capacity to achieve future value) are impossible without better targeting of support. Both sustainability and resilience are elaborated in the chapter as fairly contested and having different meanings in different contexts.

The context of this research reported in this chapter is women's tourism entrepreneurship in Sub-Saharan Africa. COVID-19, alongside the preceding Ebola outbreak (2013–2016), constitutes deadly crisis. As pandemics, both crises resulted in deaths partly due to weak health systems that delayed medical responses and cultural practices towards medical interventions and spill-over effects on economic activities in the tourism sector (Kimbu et al., 2023; Novelli et al., 2018). The limited research reveals gendered challenges pre- and post-COVID (e.g., reproductive health, financing, family role expectation, perceived male domination), in terms of how the negative impact of pandemics on women entrepreneurs far outweigh male entrepreneurs (IFC, 2021; Ukala et al., 2020; Onoshakpor et al., 2020; GEM, 2022; World Bank, 2020). The same research argues the case for women entrepreneurs to receive targeted support, whereby the effectiveness of support depends on the uptake by women entrepreneurs. What gets in the way of uptake include instances where support is not designed with the specific needs of women tourism entrepreneurs and gendered challenges preventing women entrepreneurs from

DOI: 10.4324/9781003315858-9

taking advantage of supports. Yet, the nature and rationales for decisions during a pandemic matter because a pandemic reflects a "a crucial time in which a decisive change is impending," requiring major human capital decisions and investments (Dushnitsky et al., 2020, p. 537).

Entrepreneurial responses – the mindset, attitude, behaviors, skills, and resourcefulness – of women entrepreneurs (Welter et al., 2018) are critical for the resilience of small businesses in the tourism industry against the vulnerability (Ngoasong & Kimbu, 2016) from exposure to the pandemic (Onoshakpor et al., 2020). Rather than a conventional view of resilience as the capability to bounce back, this chapter considers a dilemma underpinning entrepreneurial responses of women entrepreneurs: recover by bouncing back to the old normal versus recover by bouncing forward to the new normal? In addressing this question, this chapter uncovers that the support for women entrepreneurs enables resilience that is strategic and substantive in renewing gendered approach towards responing to crisis through tourism entrepreneurship.

This chapter therefore develops and critically explores a theoretical framework for addressing the questions around whether support for women-owned businesses simply enables a return to their previous state (bounce back) or create new visions for the future of women's entrepreneurship within the business realm (bounce forward). Creating new visions is viewed here as a responsible for those providing support as well as individual women entrepreneurs acting on the support that is being provided. The insights gained can inform improvements to existing support that incorporates learnings from women entrepreneurs who effectively navigated COVID-19 and continue to operate thriving businesses (Ukala et al., 2020). The theoretical perspectives integrate a gender-lens (G) – interventions that advance gender equality (Ngoasong & Lamptey, 2020; Quinlan & VanderBrug, 2017) and, as defined below, the context-intervention-mechanism-outcome (CIMO) logic methodology for studying resilience (Denyer et al., 2008; Bailey & Breslin, 2021).

Through analysis of primary and secondary evidence on women entrepreneurs' responses to the COVID-19 pandemic in Sub-Saharan Africa, this chapter proposes a 4G-CIMO framework for understanding how supporting women entrepreneurs enables resilience. The framework depicts the gendered context (G-C) for identifying where (i) the pandemic put at risk the sustainability of women-owned businesses, (ii) the gendered interventions (G-I) through support targeted at helping women entrepreneurs to develop resilience, (iii) the gendered mechanisms (G-M) that underpin implementation of institutional strategies for resilience, and (iv) the gendered outcomes (G-O) as resilience in women-owned businesses. Gendered is used to signify roles, behaviors, and actions that recognize the experience and biases that women entrepreneurs face. This chapter uncovers the risks, possibilities, and approaches to achieving sustainability of women-owned businesses in times of crisis. The next section provides the conceptual foundations followed by the qualitative research methodology. Next, a presentation of the findings, followed by a discussion and conclusions, are presented.

Literature Review

The conceptual foundations of this chapter cut across four inter-related concepts: sustainability, resilience, liminality, and gender-lens principles. The sustainability is the value women-owned businesses created by addressing the 3P in their strategies: (1) developing people (employability), (2) generating profits (economy), and (3) shaping the future of our planet (environment) (John, Coetsee & Flood, 2022). Drawing on Hillmann and Guenther (2021), resilience is the capacity of women-owned businesses to achieve sustainability. This definition views resilience as an organization's ability to survive and prosper owing to anticipation, preparation, response and adaptation to incremental change, and sudden disruptions over time (Kantur & İşeri-Say, 2012). At the individual level, self-resilience during a crisis requires self-belief in own capabilities, guidance/mentoring to over challenges (Matharu & Juneja, 2021), and pursuing strategies that are more sustainability oriented rather than economic oriented (Kutlu & Ngoasong, 2024). Traditionally, the resilience construct has focused on bouncing back to the old normal. However, this traditional view of resilience is limiting under the conditions of liminality – the conditions where old systems (rules, regulations, and procedures) have become futile, whereas the new ones are not yet in place (Bailey & Breslin, 2021).

An example of a changing system in Sub-Saharan Africa is the growing recognition of gender-lens investing as a distinctive category of small business financing. Gender-lens investing seeks to address long-standing underinvestment in women entrepreneurs, business risks (associated with sexual harassment, low staff retention, and value chain abuses), and limited transparency of gender equality practices in existing financing (Ngoasong & Lamptey, 2020). To this end, an alternative view of resilience is needed for addressing the dilemma women-owned businesses face while coping with the pandemic: either to bounce back or to bounce forward. This implies that, instead of aiming to return to the old ways of working (bouncing back), women entrepreneurs may opt for transforming to achieve future value (bouncing forward). Considering that the COVID-19 pandemic disproportionately affected women-owned businesses (GEM, 2021; Manolova et al., 2020), a gender lens ensures that women-owned businesses are not at risk of worsening due to existing inequalities. Sustainability and therefore resilience are impossible without fulfilling clear principles appropriate access to equal opportunities (equality) and voice (inclusion) and agency (action) by women entrepreneurs.

Denyer et al. (2008) propose a logic for interrogating prior research by focusing on how and which mechanisms (M) enable various interventions (I) leading to the desired outcomes (O) in a particular context (C). The CIMO logic synthesis centers on the four key questions: "Where?" "What?" "How?" and "Why?" (John & Vicente, 2020). The D-C (dynamic context) element investigates *where* the crisis has brought about challenges for women entrepreneurs that put their businesses at risk of achieving sustainability. Then, the I (interventions) element explores *what* institutions do to support women entrepreneurs to address the challenges suggested in the C element. This chapter emphasizes gender-lens strategies as helping to address these challenges. In the M (mechanisms) element, we explore *how* these strategies are enabled. For example, we are interested in understanding processes, procedures, and ways of working which re-enforce the gender-lens strategic decisions. Finally, the O (outcomes) element helps to explain *why* higher education institutions need the capacity to achieve their value – resilience. Our emphasis is on the performance implications of the resilience.

In terms of context, following the COVID-19 pandemic outbreak and subsequent measures (e.g., lockdowns and social distancing), we need to determine whether and how gendered challenges, such as care responsibilities and access to financing, have aggregated and potentially putting at risk the future value (the 3P dimensions of sustainability) of women-owned businesses. The gender-lens perspective reveals gendered challenges that manifest in employability, economic, and ethical leadership (E^3) necessary to develop resilience (Figure 9.1). During a crisis individuals struggle to access available support needed to develop critical skills necessary to respond (Ngoasong & Groves, 2016). Likewise, the effects of a crisis on survival and growth pose economic challenges for both early-stage and established businesses (Kutlu & Ngoasong, 2024). Building upon an ethical situationist perspective on work-life balance and dealing with uncertainty (De Clercq & Brieger, 2022), the reputation implications indicate ethical challenges for women entrepreneurs who struggle to make "trade-off decisions" to live up to their role model status in society (Kaptein, 2019, p. 1139).

Interventions are strategies aimed at helping women-owned businesses to enhance the capacity of their businesses to achieve sustainability. Resilience strategies include making use of resourcefulness (Welter et al., 2018), improvising in new markets and partnership (Ngoasong & Kimbu, 2016), and developing organizational guidance to support employees and business operations (Ngoasong & Groves, 2016; Matharu & Juneja, 2021). Van Der Vegt et al. (2015) argues that the intra-organization resilience in small- and medium-sized businesses comes from the capacity to achieve sustainability by coordinating the various internal units. By contrast, the inter-organizational resilience as the capacity to achieve sustainability is about leveraging power and resource dependencies with external institutions.

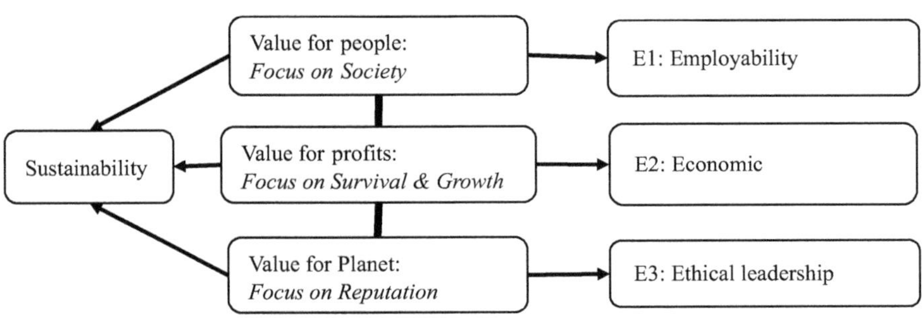

Figure 9.1 A E^3 framework for gendered challenges to sustainability

Mechanism allows for analysis from the macro-level to the micro-level (Raub & Voss, 2017), processes and behaviors that may support the formation of new strategic and organizational capabilities in women-owned businesses. Developing resilience requires resources and capabilities, the creation of which involves substantial investments of time, effort, and finance; yet does resilience, as capacity to achieve sustainability, add value? (Hillmann & Guenther, 2021). Adding value to entrepreneurial performance is a concern for women entrepreneurs and those institutions that support women's entrepreneurship are likely to express before taking on board the recommendations offered in this chapter about where, what, and how to address post-pandemic challenges. Thus, understanding of mechanism is critical for effective implementation of support for women-owned businesses in a particular context.

Methods

This chapter combines qualitative content analysis of data collected from secondary (published) and primary (interviews) sources, an extensive search and review of published accounts of gendered challenges, entrepreneurial responses, and types of support for women entrepreneurs in African countries. Sources include UN Women, World Bank, Global Entrepreneurship Monitor, International Finance Corporation, Mastercard Foundation, and bilateral development agencies. The review captures empirical, agenda, and evaluation accounts that offer context-relevant evidence of the lived experiences of women entrepreneurs regarding COVID-19 (Frimpong et al., 2022). Overall, the evidence provides a range of approaches through which financial and non-financial support for women entrepreneurs during crises increases their capacity to achieve sustainability. Primary data collected from in-depth qualitative case studies developed through semi-structured interviews on women-owned businesses. The case studies were part of a larger dataset collected pre- and post-COVID by the author when authoring a research monograph on women's entrepreneurial journeys in Africa (Ngoasong, 2023).

The empirical evidence contextualizes country-specific examples. The published cases allow for a more in-depth exploration of the issues and provide useful sources for the purpose of this study. Data analysis proceeded through participants' accounts (first order), second-order codes, and aggregate theoretical dimensions (Gioia et al., 2013). Participants' accounts are key words or phrases within entire responses captured in transcripts and fieldnotes and are contents analyzed to create two data analytic structures (Figures 9.2 and 9.3), each revealing first-order, second-order, and aggregate theme for theorizing how support for women-owned businesses enables the development of effective strategies for resilience – the capacity to achieve sustainability in business. For instance, to uncover the support for women entrepreneurs during the pandemic (Figure 9.2), women entrepreneurs' experiences of how the pandemic affected their businesses (first order) are aggregated as crisis-related challenges (second order) that put at risk the sustainability of women-owned business (aggregate theme). To address these challenges, women entrepreneurs sought support (first-order), grouped within a gender-lens perspective (second order), which enables resilience (aggregate). A similar analytic procedure was used to uncover how the strategies for resilience produce performance implications for resilience (Figure 9.3).

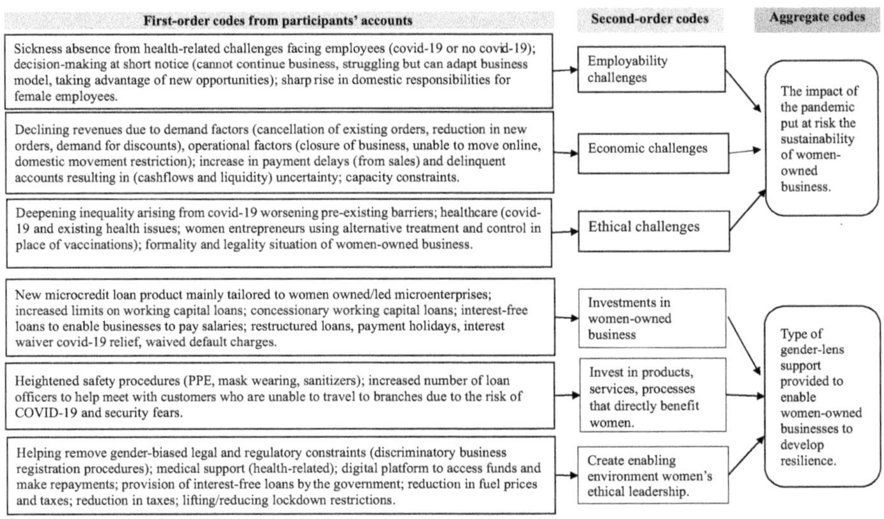

Figure 9.2 Data analytic framework for resilience strategies resulting from mechanisms for support

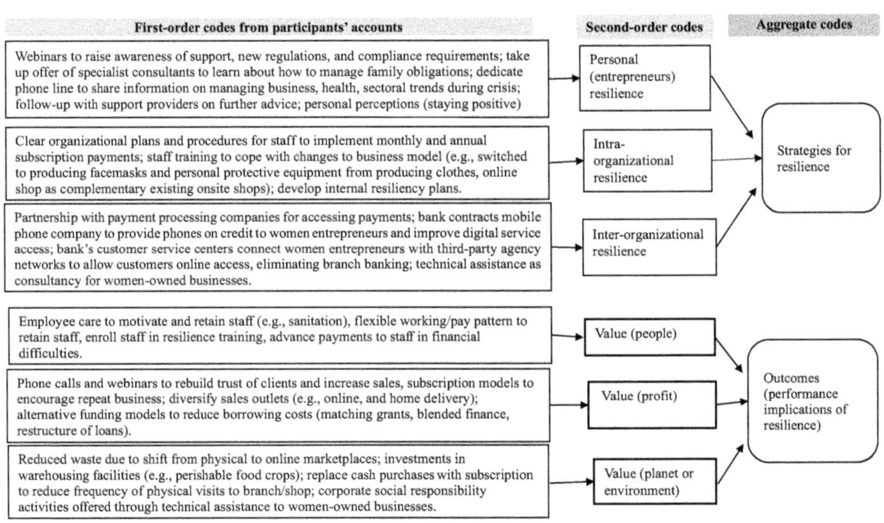

Figure 9.3 Data analytic framework for uncovering types of support for women entrepreneurs in times of crisis

Findings

Why Support for Women Entrepreneurs in an (Un)equal Pandemic Environment?

Overall, the evidence shows common themes related to personal (woman entrepreneur), intra-organizational (within their business), and inter-organizational (relationships within domestic and global value chains) dimensions of employability, economic, and ethical leadership challenges that put at risk the sustainability of women-owned business. Admittedly, some challenges affected both businesses owned by women and those owned by men – uncertainty around production sales, fall in income due to reduced consumer demand, and supply chain constraints arising from government lock-down measures. As captured below, it becomes important to focus on the evidence relating to women, which reveals the gendered nature of the unequal impacts arising from COVID-19:

> More than 80% of women business leaders in Kenya reported that domestic responsibilities were having a significant impact on their work. … Over 50% of women business leaders reported struggling to pay household bills.
>
> [GEM, 2022]

The references to domestic responsibilities and the struggle to pay household bills are related to gender roles in the African family. Whereas men are typically seen as breadwinners for the family, women are known to perform domestic duties, with many women entrepreneurs combining their business and domestic duties. Evidence came from women entrepreneurs who spoke about "being a single mum with no single support for childcare from authorities" (Entrepreneur 11, Cameroon), "There are people in my husband's family and even society at large who think I should be at home, not going around running a business" (Entrepreneur 4, Burkina Fasso), and "Running this business part-time allows me to look after my children while generating income doing what am passionate about" (Entrepreneur 7, Kenya).

However, the data reveal support for women entrepreneurs during COVID-19 as (i) training, mentoring, and networking; (ii) financing; and (iii) creating or improving the enabling environment for women's entrepreneurship. Women entrepreneurs took own initiative to deal with whatever risks they encountered during the pandemic, including within the business realm. Examples in Cameroon included a "reduction and variations in working hours, salaries, payment; cancellation of supplier orders, postponing investments and/or drawing on savings," actions that constitute a view of resilience as capacity to bounce back from a crisis in the short term (Kamga & Deffo, 2022, p. 885). While these measures are essential for the business, many women entrepreneurs acknowledged that they do not necessarily address the domestic responsibilities and perceptions of gender roles in society.

As an illustration, consider Entrepreneur 1, CEO of a Nigeria-based natural skincare company that sources shear, obtained from the nuts of shea trees and exported for use in producing premium skin care products and for food preparation in the United Kingdom. Though this is a manufacturing business, it plays a complementary role in the tourism industry (Kutlu & Ngoasong, 2024) in the provision of complementary products and services that contribute to providing a holistic tourism experience. This is in the sense that visitors to the farms in Nigeria relate to rural tourism (Ohe & Kurihara, 2013) and visitors to premium high-street brand in cities (e.g. Weidenfeld et al., 2011). During 2019–2021, the company constructed a manufacturing facility in Nigeria, employed 22 women, working full time to produce shear butter from natural plants. When asked to describe how COVID-19 put businesses at risk, Entrepreneur 1 spoke of health of workers (employability), poverty and disruption supply (economic), and changing cultural practices (ethical leadership) challenges:

> With Covid-19, you can do social distancing and hand sensing. But the people were thinking more about 'how I am going to make money today,' not much about the future of their health. Remember these are poor families. Men were cutting the shear trees for their wives to use as wood in cooking during the collection season when we needed the collect the shear. We had to focus on support the women who will intend support their families. Trying to change that mindset is incredibly challenging.
>
> (Interview, Entrepreneur 1)

During the pandemic, women entrepreneurs relied on foodstuffs and therapies to manage COVID-19, in response to limited access to vaccines and cultural practices in which the population is skeptical about vaccinations. Evidence from Ghana and South Africa identified the perception that local vegetables, ginger, garlic, fruits, lemon, food spices, herbal medicine, hot food, and warm water can boost the immune system and kill the COVID-19 virus (Frimpong et al., 2022). The outcomes are the value to the people in terms of the psychological boost arising from the knowledge (Byrne & Shepherd, 2015) that local alternatives assured their personal safety. However, there are ethical challenges surrounding the effectiveness of beliefs that foodstuffs or fruits, as against medical interventions (vaccinations, and access to these) for resilience (Ngoasong & Groves, 2016). This is another area where support for women entrepreneurs can help to build their resilience. Examples in the data include heightened safety procedures through providing personal protective equipment, mask wearing, sanitizers, and raising awareness about the importance of vaccinations (e.g., Wi-Fi, 2020).

Gender-Lens Mechanisms for Enabling Strategies for Resilience

To understand how the mechanisms for supporting women-owned businesses to build resilience recognize and deal with the gendered nature of the impact of a crisis, this section focuses on how the processes and procedures used to deliver support reflect the three gender-lens strategies (Quinlan & VanderBrug, 2017): investing in women-owned business; investing in products, services, and processes that directly benefit women; and investing in developing women's ethical leadership in business.

Consider investing in women-owned business through an explicit gender-lens strategy in which financial institutions create a dedicated fund for providing direct financing for women-owned businesses (Ngoasong & Lamptey, 2020). An example of a gender-lens investment as enabling resilience during COVID-19 is captured in the following quote:

> A Kenyan bank brought in specialist consultants to facilitate conversations with women about managing family obligations in Kenya. … Specialized business training and grant-matching assistance to continue to operate in Burkina Faso and Nigeria. … Women insurance program protected women-owned businesses against the risks, and better control of assets, in Cameroon and Ghana. Financial products/services tailored to women-led cooperatives and to train WSMEs, large buyers of fresh produce, and financial institutions on asset-based lending in Côte d'Ivoire.
>
> (Wi-Fi, 2020)

Second, investing in products, services, and processes that directly benefit women (Quinlan & VanderBrug, 2017). According to GEM (2022), "both women and men cite family care as critical measure for business relief, with the highest average rates reported in Nigeria (49%)." To understand the importance of investing in products for women in times of crisis, consider the case of Entrepreneur 2, CEO, and Founder of a social enterprise in Kenya that provides sanitary towels to women (families) and young girls (in secondary and high schools) from the poorest backgrounds. Funding for this enterprise partly comes from sales donations locally and sales from books authored through the work of the organization. Entrepreneur 3 described how the pandemic put at risk the sustainability of her business, as follows:

> Access to this essential product for women worsened during covid-19 because of the lockdown measures. There was no way for us to distribute the sanitary products in schools. We normally met the girls on campus. And for women in rural communities the costs of hiring vehicles to reach them was just too high. We were receiving phone calls and text messages from women asking for help. Imagine the thought of not being able to reach them.
>
> (Entrepreneur 2, Nigeria)

In January 2021, Entrepreneur 3 enrolled into a competitively run virtual accelerator that was offering cash support (to cover expenses), mentorship and leadership develop training over a six-month period. The cash support covered mobile phone bills for volunteers involved in raising funds from local organizations and travel expenses to distribute sanitary products to local centers that was within reach for the women. Here, we see personal resilience strategies at play, in terms of the entrepreneur taking personal initiative and being resourceful in responding to a crisis (Welter et al., 2018). The support received through the accelerator enables resilience in the short term, in this case during the six months acceleration period.

Third, creating enabling environment that supports women's ethical leadership in business. Here, resilience goes beyond the personal and organizational focus to emphasize the inter-organizational. Public-private partnerships involving governmental and non-governmental stakeholders supporting women's representation in decision-making and other stakeholders as observers, advisers, and/or donors help to bring resilience in action. Examples include the Women Entrepreneurs Finance Initiative that seeks "to amplify the resilience of women entrepreneurs" and which raised "almost $50 million to addresses the needs of women entrepreneurs that have arisen due to the COVID-19 crisis, by encouraging innovation, digital transformation and developing partnerships to facilitate greater access to financing for women-led businesses" (Wi-Fi, 2020). This gender-lens solution includes lobbying governments to address legal and regulatory requirements that are perceived as gender bias, such as "childcare support, and ethnic practices that favor men in family succession planning" (Entrepreneur 8, Ghana).

To understand the performance implications of a gender-lens mechanism for enabling resilience that enhances the capacity for sustainability in women-owned businesses, consider the case of Entrepreneur 3, who grows sells retail fresh food to hospitality businesses in Ghana but who only succeeded in securing external financing after more than ten years of trying.

> For over ten years, I was not successful in getting a loan from a bank because as a woman I am not entitled to a property. … I don't have a husband to be guarantor. So, there is no collateral security. … Then the interest rates are too high.
>
> (Entrepreneur 3, Ghana)

The mention of high interest rate above is a long-standing obstacle that women entrepreneurs face and contradicts the view that simply making it easier for women entrepreneurs to access loans can build the resilience of their business. For Entrepreneur 3, the solution investing with a gender lens, meaning intentionally investing in women-owned business, with an objective to make profits while creating measurable impact on society (Ngoasong & Lamptey, 2020). In addition to equity finance, the business also benefited from technical support, reflecting the need to build resilience to macro-level challenges (Raub & Voss, 2017): "technical assistance facility to procure a new accounting software system and to pay for an experienced retired executive to come in and strategize and collaborate with my team to implement a new accounting framework" (Entrepreneur 3). This is related to the view of resilience as the capacity to bounce back from adversity as essential for achieving sustainability (Hillmann & Guenther, 2021). The performance implications for Entrepreneur 3 because of gender-lens investing was (i) increased total sales, (ii) growing and diverse customers and sourcing channels, (iii) improved cost-competitiveness, and (iv) a reliable supply of quality raw materials and sourcing of food crops.

Discussion

4G-CIMO Framework for Enabling Gender-Lens Resilience in Women-Owned Businesses

This chapter has demonstrated how enabling women-owned business to achieve and maintain effective strategies for resilience can enhance their capacity to achieve sustainability. Based on the findings, a 4G-CIMO framework is proposed (Figure 9.4) that reconceptualizes the CIMO logic developed in Denyer et al. (2008) by introducing a gender-lens perspective (Ngoasong & Lamptey, 2020; Quinlan & VanderBrug, 2017) to each of four key questions of Context (Where?), Intervention (What?), Mechanism (How?), and Outcome (Why?). In this reconceptualization, understanding how support for women-owned businesses can enable resilience focuses on why and how gender-lens mechanisms (G-M) enable various interventions by women entrepreneurs (G-I) leading to desired outcomes (G-O) for their businesses in each gendered context (G-C). The use of gendered emphasizes the importance of a gender-lens view at macro-levels (government policy) and micro-levels (individual business support). The framework provides a guide for women entrepreneurs (and those who provide support to them) interested in pursuing sustainability.

The framework starts with the gendered contexts (G-C) for understanding how a pandemic affects women entrepreneurs, puts at risks the sustainability of their business, and provides necessity support to enable resilience to the pandemic. Employability challenges include staff illness and knowledge and skills gap about

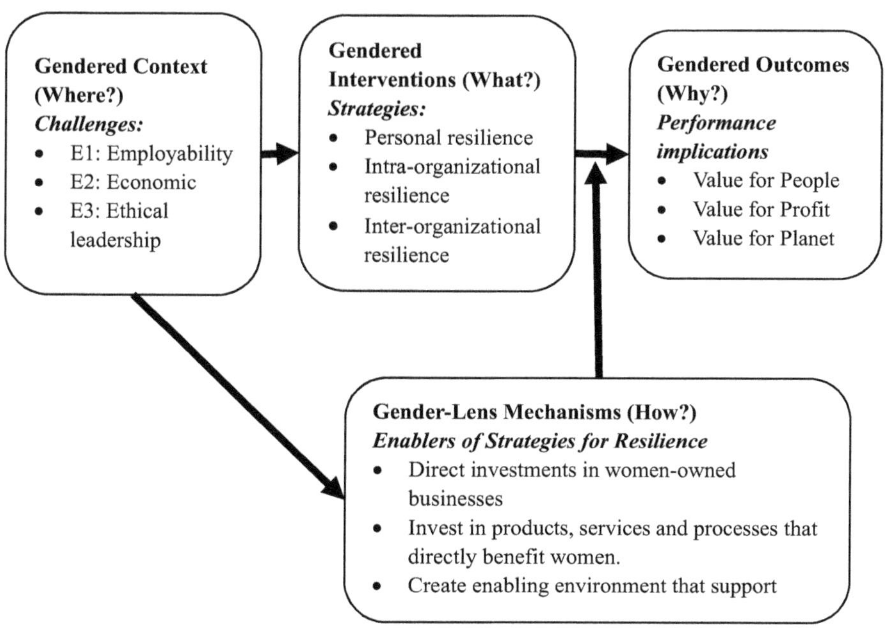

Figure 9.4 4G-CIMO framework for enabling resilience in women-owned businesses

how to respond to a pandemic, while economic challenges include financing and revenues/incomes from sales. There are gendered challenges that arise from societal perceptions of men versus women in managing relationships with customers and suppliers (fear of failure). Whereas the COVID-19 pandemic affected all businesses, the impacts do not equally, with businesses owned by women seen as more negatively impacted compared to those owned by men (Wi-Fi, 2020). This is in addition to long-standing perceptions of women as embodying the social role of caregiver as mothers, wives, or stay-at-home spouses, which creates extra pressure on women entrepreneurs. The findings in the chapter reveal that the resulting inequality in the impacts of a pandemic creates an ethical leadership challenge for women entrepreneurs.

The above challenges call for gender-lens mechanisms (G-M) for supporting women-owned businesses in time of crisis. Ngoasong and Lamptey (2020) distinguish between implicit and explicit gender-lens strategies for supporting women-owned businesses. During the COVID-19 pandemic, most financial institutions implemented implicit gender-lens strategies, as evidenced in data that do not disaggregate loan applications that came from a male versus female entrepreneurs (IFC). And example of how an implicit gender-lens strategy does not look at the gender identity of entrepreneurs is where loan offices focus on the finance sections of loan applications, rather than the personal circumstances of the entrepreneur applicant in approval/rejection decisions (Ngoasong & Lamptey, 2020). Other institutions implemented a gender-lens approach that explicitly audits evidence of support for women entrepreneurs during COVID-19 (Wi-Fi, 2020). Approaches that draw conclusions from survey of both men and women in equal measure risk ignoring the gendered challenges discussed in this chapter and perpetuating gender biases that put at risks the sustainability of women-owned business. A gender-lens strategy ensures that decision-making processes reflect greater concerns about the unique challenges facing women-owned businesses in achieving sustainability.

Gendered challenges may exist, and gender-lens mechanisms may be available to supporting women entrepreneurs and thereby enable them to develop resilience strategies. Awareness and usage of support by women entrepreneurs is the only way to ensure resilience. That is why women entrepreneurs are urged to take advantage of opportunities created by COVID-19 (Manolova et al., 2020). This is related to a gendered intervention (G-I) for building personal, intra-organizational, and inter-organizational resilience. The personal emphasizes women entrepreneurs' motivations – "statuses, both social and financial, of being a woman owner-manager (self), enterprise survival and sustainability and serving defined community needs through business activities" (Kimbu & Ngoasong, 2016). Due to a pandemic, women made tough decisions internally, with staff. Decisions require buy-in from staff to avoid the likelihood of resistance to the implementation of those actions. Gender-lens strategies prioritize gender awareness in listening to staff and customers, seizing opportunity to gain experience, and exploring new ways of doing business (Wi-Fi, 2020). At inter-organizational resilience, relationships with financial institutions, markets, and employability (Conz et al., 2017) create spaces for women to confidently function within the company.

Finally, gendered outcomes (G-O) refer to the performance implications or the value derived from resilience, for people, profit, and planet. The people element includes the welfare/well-being of employees and the ethical leadership. There are emotional worries that come from fear of failure for a business that is not making profit (Byrne & Shepherd, 2015). This matters more for women entrepreneurs because, compared to male entrepreneurs, women cite fear of failure, lack of self-confidence, and everyday challenges as preventing them from seeking support to start and grow businesses (Carranza et al., 2018). The profit element responds to economic challenges (finance, sales, operations, revenue/income/profit), while planet relates to the environment. What we know about support for women entrepreneurs during COVID-19, whether on African countries (Kamga & Deffo, 2022) or other parts of the world (Matharu & Juneja, 2021; GEM, 2022) has been about resilience strategies to bounce back post-pandemic. Survival requires addressing immediate challenges; adaptation requires reallocating resources, while innovation emphasizes long-term strategic shifts in ways of doing businesses. It is possible that gender-lens resilience may enable women entrepreneurs to realize sustainable business models – business models which allow aligning and creating synergies between the performance expectations of resilience and the sustainable development goals (Manolova et al., 2020). Future research should track the process that women entrepreneurs go through from securing financial and non-financial support, developing, and implementing resilience strategies and achieving sustainability.

Yet, it also follows from the 4G-CIMO framework that, despite its value, focusing on sustainability is not risk-free. The value of sustainability (3Ps) may depend on the severity of a combination of employability, economic, and ethical leadership challenges, as demonstrated earlier by the E3 Framework (Figure 9.1). For instance, the profitability (economic challenges) of women-owned businesses often comes under pressure because of emerging societal challenges that create conflicts of interest (e.g., immediate concerns for poverty alleviation versus concerns about profit growth and environment). Similarly, changes in the perspectives among policy makers can affect what constitute an enabling environment for different industry sectors, which may not always favor women-owned businesses. Driven by economistic and/or reputational concerns, public and private institutions successfully extending access to finance and training for women entrepreneurs must go beyond short-term economistic and capacity development considerations towards socially and environmentally sustainable considerations.

For women-owned digital businesses, their processes had already been adapted to virtual ways of working before the pandemic's outbreak, which helped to offset the pandemic impacts on business (GEM, 2022). Despite the development of digital technology infrastructure to support access to the internet and online buy-sell transactions, access to online payment capabilities and local transport and distribution of goods to end users remains major challenges. Paying for third-party online platforms helps women-owned businesses to reach customers without the time-consuming efforts for them to travel to bank branches or shops (IFC, 2021) is an environmentally sustainable solution because of reduced carbon emission for physical travel (John et al., 2022). Admittedly, such support risk serving as reactive measures to a crisis that may not produce sustainable outcomes where online payment capabilities do not reach the full breadth of women entrepreneurs' target market. This is an area in need of future research.

Conclusions

This chapter proposes a 4G-CIMO framework for understanding and pursuing an alternative view of resilience, based on an investigation of African women's entrepreneurial reactions to the COVID-19 pandemic and considering how. This chapter also uncovers the extent to which variations in women's business activities pre- and post-pandemic empower these businesses to attain sustainability. Women may choose to bounce forward and embrace business transformation through innovation, improvisation, and diversification rather than bounce back to the status quo in response to changes brought on by a pandemic. Institutions that support women entrepreneurs should equally embrace this sustainability perspective within gender-lens policies, programs, and initiatives. This will ensure that support enables strategies for resilience that produce intended performance outcomes.

References

Bailey, K., & Breslin, D. (2021). The COVID-19 pandemic: What can we learn from past research in organizations and management? *International Journal of Management Reviews*, *23*(1), 3–6.

Byrne, O., & Shepherd, D.A. (2015). Different strokes for different folks: Entrepreneurial narratives of emotion, cognition, and making sense of business failure. *Entrepreneurship Theory and Practice*, *39*(2), 375–405.

Carranza, E., Dhakal, C., & Love, I. (2018). *Female Entrepreneurs: How and Why Are They Different?* Washington, DC: The World Bank. https://documents1.worldbank.org/curated/en/400121542883319809/pdf/Female-Entrepreneurs-How-and-Why-are-They-Different.pdf

Conz, E., Denicolai, S., & Zucchella, A. (2017). The resilience strategies of SMEs in mature clusters. *Journal of Enterprising Communities*, *11*(1), 186–210.

De Clercq, D., & Brieger, S.A. (2022). When discrimination is worse, autonomy is key: How women entrepreneurs leverage job autonomy resources to find work-life balance. *Journal of Business Ethics*, *177*(3), 665–682.

Denyer, D., Tranfield, D., & Van Aken, J.E. (2008). Developing design propositions through research synthesis. *Organization Studies*, *29*(3), 393–413.

Dushnitsky, G., Graebner, M.E., & Zott, C. (2020). Entrepreneurial responses to crisis. *Strategic Entrepreneurship Journal*, *14*(4), 537–548.

Frimpong, E.K., Yamoah, P., Wiafe, E., Demana, P.H., & Matlala, M. (2022). Market women's perspectives on coronavirus disease 2019 (COVID-19): The case of Ghana and South Africa. *International Journal of Environmental Research in Public Health*, *19*(15), 9658.

GEM. (2021). *Global Entrepreneurship Monitor 2021/2022 Global Report: Women's Entrepreneurship, Thriving through Crises*. https://www.gemconsortium.org/reports/womens-entrepreneurship

GEM. (2022). *COVID-19 Impacts on Women Entrepreneurs in Emerging Economies: Insights and Indicators*. https://www.gemconsortium.org/images/media/covid-19-impacts-on-women-in-emerging-economies-1644427445.pdf

Gioia, D.A., Corley, K.G., & Hamilton, A.L. (2013). Seeking qualitative rigor in inductive research: Notes on the Gioia methodology. *Organizational Research Methods, 16*(1), 15–31.

Hillmann, J., & Guenther, E. (2021). Organizational resilience: A valuable construct for management research? *International Journal of Management Reviews, 23*(1), 7–44.

IFC. (2021). *COVID-19 and Women-Led MSMEs in Sub-Saharan Africa: Examining the Impact, Responses, and Solutions*. International Finance Corporation (IFC). https://www.ifc.org/content/dam/ifc/doc/mgrt/202103-covid-19-and-women-led-msmes-sub-saharan-africa.pdf

John, A., Coetsee, J., & Flood, P.C. (2022). Understanding the mechanisms of sustainable capitalism: The 4S model. *Business Ethics, the Environment and Responsibility, 32*, 15–24. https://doi.org/10.1111/beer.12495

John, A., & Vicente, G. (2020). Institutional foundations of management of natural disasters: Lessons from the recent cyclones in Mozambique. In Leite da Silva, C.A., Oliveira da Silva, E.S., & Vilar-Lopes, G. (Eds.), *Humanitarian Questions and Airforce Power* (Chapter 11, pp. 63–86). Rio de Janeiro: Luzes: Communication, Art, and Culture.

Kamga, B.F., & Deffo, R.N. (2022). Analysis of the resilience strategies of Cameroonian companies in the face of Covid-19 and their effects on activity. *Journal of International Development, 34*(4), 880–897.

Kantur, D., & İşeri-Say, A. (2012). Organizational resilience: A conceptual integrative framework. *Journal of Management and Organization, 18*(6), 762–773.

Kaptein, M. (2019). The moral entrepreneur: A new component of ethical leadership. *Journal of Business Ethics, 156*(4), 1135–1150.

Kimbu, A.N., Adam, I., Dayour, F., & De Jong, A. (2023). COVID-19-induced redundancy and socio-psychological well-being of tourism employees: Implications for organizational recovery in a resource-scarce context. *Journal of Travel Research, 62*(1), 55–74.

Kimbu, A.N., & Ngoasong, M.Z. (2016). Women as vectors of social entrepreneurship. *Annals of Tourism Research, 60*, 63–79.

Kutlu, G., & Ngoasong, M.Z. (2024). A framework for gender influences on sustainable business models in women's tourism entrepreneurship: Doing and re-doing gender. *Journal of Sustainable Tourism, 32*(3), 500–518.

Manolova, T.S., Brush, C.G., Edelman, L.F., & Elam, A. (2020). Pivoting to stay the course: How women entrepreneurs take advantage of opportunities created by the COVID-19 pandemic. *International Small Business Journal, 38*(6), 481–491.

Matharu, S. K., & Juneja, D. (2021). Factors impacting resilience of women entrepreneurs in India in the face of COVID-19. *Vision: The Journal of Business Perspective*. https://doi.org/10.1177/09722629211043299

Ngoasong, M.Z. (2023). *Women's Entrepreneurial Journeys in Sub-Saharan Africa*. Cheltenham: Edward Edgar.

Ngoasong, M.Z., & Groves, W.N. (2016). Determinants of personal resilience in the workplace: Nurse prescribing in an African work context. *Human Resource Development International, 19*(3), 229–244.

Ngoasong, M.Z., and Kimbu, A.N. (2016). Informal microfinance institutions and development-led tourism entrepreneurship. *Tourism Management, 52*, 430–439.

Ngoasong, M.Z., & Lamptey, R.O. (2020). Gender lens investing in the African context. In Sarmento, E. de Morais & Herman, R.P. (Eds.), *Global Handbook of Impact Investing: Solving Global Problems Via Smarter Capital Markets Towards a More Sustainable Society* (pp. 273–302). Hoboken, NJ: John Wiley & Sons, Inc.

Novelli, M., Burgess, L.G., Jones, A., & Ritchie, B.W. (2018). 'No Ebola…still doomed' – The Ebola-induced tourism crisis. *Annals of Tourism Research, 70*, 76–87.

Ohe, Y., & Kurihara, S. (2013). Evaluating the complementary relationship between local brand farm products and rural tourism: Evidence from Japan. *Tourism Management, 35*, 278–283.

Onoshakpor, C., Etuknwa, A., & Karamalla-Gaiballa, N. (2020). Strategic flexibility and organizational resilience of women entrepreneurs in Africa during the COVID-19 pandemic. *Research Journal of Business and Management, 7*(4), 277–287.

Quinlan, J.P., & VanderBrug, J. (2017). *Gender Lens Investing Uncovering Opportunities for Growth, Returns, and Impact*. Hoboken, NJ: John Wiley & Sons.

Raub, W., & Voss, T. (2017). Micro-macro models in sociology: Antecedents of Coleman's diagram. In Jann, B. & Przepiorka, W. (Eds.), *Social Dilemmas, Institutions, and the Evolution of Cooperation* (pp. 11–36). Berlin: De Gruyter Oldenbourg.

Ukala, E., Abril, E.R., & Esther Dassanou, E. (2020). *Transformative Policy Solutions to Support Women-Led Businesses in Africa in a Post Covid-19 World*. https://www.tralac.org/documents/resources/covid-19/regional/3980-policy-brief-transformative-policy-solutions-to-support-women-led-businesses-in-africa-in-a-post-covid-19-world-july-2020/file.html

Van Der Vegt, G.S., Essens, P., Wahlstrom, M., & George, G. (2015). From the editors: Managing risk and resilience. *Academy of Management Journal, 58*(4), 971–980.

Weidenfeld, A., Butler, R., & Williams, A.M. (2011). The role of clustering, cooperation and complementarities in the visitor attraction sector. *Current Issues in Tourism, 14*(7), 595–629.

Welter, F., Xheneti, M., & Smallbone, D. (2018). Entrepreneurial resourcefulness in unstable institutional contexts: The example of European Union borderlands. *Strategic Entrepreneurship Journal, 12*(1), 23–53.

Wi-Fi. (2020). *Annual Report: Women Entrepreneur Finance Initiatives*. https://we-fi.org/wp-content/uploads/2021/03/WeFi-Annual-Report-2020.pdf

World Bank. (2020). *Gender Dimensions of the COVID-19 Pandemic*. https://documents1.worldbank.org/curated/en/618731587147227244/pdf/Gender-Dimensions-of-the-COVID-19-Pandemic.pdf

10 Conclusion

The Imperative of Gender and Social Policy in Tourism Entrepreneurship Research

Anna de Jong and Albert Nsom Kimbu

Introduction

As part of a neoliberal focus on growth and employment generation, many countries have developed specific policy interventions to support tourism entrepreneurs. Such interventions have tended to be economic in their prioritisation, designed to address 'market failures', and target general support for entrepreneurship (Ahl and Nelson, 2015). Such generalist support has been found to fail in their objective of achieving growth because they have tended to not engage with the gendered complexity of tourism entrepreneurship, whereby support is generally determined by masculine and Westernised (mis)conceptions that assume 'women' lack the required skills and capital to generate high growth (Figueroa-Domecq et al., 2024). This construction of women entrepreneurs as vulnerable and requiring support reinforces gendered divisions without attempting to genuinely understand the structural governance forces that produce such divisions to begin with (Figueroa-Domecq, de Jong and Williams, 2020).

Importantly, however, as governments respond to, and reflect upon, contemporary global challenges, such as climate change and the Covid-19 pandemic, the role of *social* policy in supporting entrepreneurs and the structural barriers that inhibit an ability to do so have become increasingly recognised (Kimbu, Ngoasong and de Jong, 2024). Contemporary global challenges have brought to light the ways through which certain entrepreneurial identities are disproportionality affected. The direct and indirect effects of the Covid 19 pandemic, for example, have been felt most acutely by those in the Global South, where most tourism enterprises are micro, small and medium sized (MSMEs) and are owned by women and members of disadvantaged groups and communities. As a result, significant numbers of tourism enterprises have closed, leading to millions of job losses and financial hardship (Rogerson and Baum, 2020; UNWTO, 2020). Conversely, regions whereby government support was higher and more targeted were understood to experience lower levels of enterprise closures and financial hardship (UNWTO, 2020). Such recent events lay bare the need for greater interrogation of the role, and potential, of social policy within the context of entrepreneurship generally and gender and entrepreneurship specifically.

DOI: 10.4324/9781003315858-10

An emerging body of literature exists in peer-reviewed journals that seek to address the limited interrogation of social policy within the context of gender and tourism entrepreneurship (cf. the recent special issue on this topic in the *Journal of Sustainable Tourism*, edited by Kimbu, Ngoasong and de Jong (2024)). Yet, applying a social lens to considerations of gender and entrepreneurship remains in its infancy within an area of literature that has traditionally been dominated by essentialist and universalist constructions of entrepreneurs as masculine, white and from the Global North, and has as such only relatively recently become attuned to the gendered dimensions of entrepreneurial performance (Figueroa-Domecq, de Jong and Williams, 2020; Hmieleski and Sheppard, 2019) – with even less attention given to the ways social policy interventions might exacerbate or alleviate such inequities. Hence, the impetus for this edited collection is that we have hoped to play a role in generating a research agenda that introduces social policy considerations to the interrogation of gender and entrepreneurship.

Researching Gender, Tourism Entrepreneurship and Social Policy

This collection had the aspiration to uncover and evidence research and praxis unpacking the complexity of gendered entrepreneurial pathways within the context of tourism and the role/impact of social policy in shaping this. However, putting together the collection was fraught with some challenges due to several issues. The extant metrics-driven culture encourages and requires researchers to target publishing mainly in specific 'high ranking' journals that would ensure their career progression and also judging paper quality by where it was published. This made it challenging to get contributions from colleagues, especially early and mid-career researchers, universities and/or countries in the Global North in spite of interest in contributing to the collection from potential contributors. Relatedly, in spite of interest in the book, time constraints were identified by many potential contributors as a deterring factor in submissions as they preferred to focus on the core demands and expectations of their institutions, most of which did not consider book chapters in the annual academic appraisal process.

It is also worth noting that the neoliberal approach emphasising the importance of metrics in publications is perpetuating and reinforcing the already pervasive structural inequalities existing in academia and academics not only between different world regions but also between English speaking and non-English speaking academics and importantly, as well, between genders wherein women academics (and academics from marginalised groups) tend to be most affected experiencing greater obstacles (such as time and caring responsibilities) in their career journeys (Crook, 2020). With most of the highly ranked tourism and entrepreneurship journals being published only in the English language, non-native speakers are often at a disadvantage as they tend to follow agendas and standards set by English-speaking scholars. This further widens the already existing chasm in the production of knowledge that is deemed to be acceptable and thereby indirectly discrediting alternative, liberal and creative knowledge production pathways that emanate from researchers in and/or from non-Western contexts (Yang and Ong, 2020).

In due consideration of the above, in this book, we adopted a developmental approach in the reviewing and editorial process, providing multiple rounds of feedback and guidance to the contributors on how they could strengthen different aspects of their submissions. This notwithstanding, a few submissions were regrettably not included in the final collection. This was mainly due to the tokenistic integration and use of 'gender' and/or 'social policy' in these submissions. There is a growing recognition and consideration of understanding the importance of gender, entrepreneurship and (social) policy in tourism research. This is primarily due to the increasing recognition by organisations such as the UN Tourism the critical role tourism entrepreneurship undertakes, if properly deployed, in reducing poverty (UN SDG1), fostering gender equality (UN SDG5) and empowering women and other marginalised groups and thereby reducing inequalities (UN SDG10). This notwithstanding, in putting together this collection and from our experiences as journal reviewers, the tendency in many hospitality and tourism publications, especially those from Global South destinations, has been to adopt the 'add women and stir' approach employing mainly qualitative methods of data collection, and a focus on providing descriptive comparisons of 'gender differences' (Henderson, 1994), with only very limited reflections seeking to unpack the complexities of gender relations.

Relatedly, most tourism policy-related research is centred on unpacking and addressing economic policy challenges over social policies, and even when social policy is mentioned, it is often only discussed in passing in the conclusion and implications section without any clear reflections of the gendered nature of such policies. In commending the slow but steady and growing attention to research on gender and entrepreneurship as whole, more still needs to be undertaken to understand and integrate the social policy dimension especially in relation to inclusion and equity. Furthermore, notwithstanding the 'first mover' advantage and near monopoly in determining the research foci and the production of related knowledge enjoyed by the Global North, it is hoped that this and similar extant (see e.g., Swart et al., 2024) and future research will help in reducing/bridging the gap through the mentoring and developmental approach adopted in the editorial process.

In pondering on this volume and the whole editorial journey, it is worth noting that despite a clear indication in our call for chapters and the adoption of a targeted approach in canvassing for submissions, of all the accepted submissions, only one chapter (7) included men and/or covered the subject of masculinity. This clearly indicates the ongoing (mis)assumption within academia and industry that gender equates to women. More research is therefore needed to foster awareness and wider understanding of gender within a broader spectrum underpinned by an intersectional lens if gender-based inequalities are to be understood and addressed within the context of tourism entrepreneurship.

The Social Policy Imperative for Gender and Tourism Entrepreneurship

Given the relatively recent recognition regarding the importance of social policy in supporting entrepreneurship and that there is still much to be examined regarding the ways women might benefit through such government backing, several chapters in this edited collection attempted to make a stronger case in this direction. Andrés Morales Pachón and Sara Calvo Martinez in Chapter 8, for example, undertook a broad perspective in examining Colombian ethnotourism initiatives informed through the lens of Buen Vivir; Buen Vivir being a comprehensive development approach centred on community solidarity, reciprocity and a balanced coexistence with nature. The chapter is valuable in arguing for a redirection in the type of support provided by governments, away from a prioritisation with economic growth and towards a more diversified approach that attempts to foster social bonds and cultural pride in ways that lead to a more sustainable economy.

Taking a specific focus with women in the Pervian tourism industry, in Chapter 2, Del-Aguila-Arcentales, Alvarez-Risco and Yañez examined the significant efforts put in place to encourage women to take up entrepreneurial activities. This chapter highlighted the imperative in ensuring sufficient social policy support is in place for those women who do respond to efforts to attract them to entrepreneurship. The authors note that in Peru, policy interventions have included enhanced access to finance, digital education and mentoring networks. Mugendi and Kiarie built on these ideas in Chapter 6, in their examination of women entrepreneurs in the Kenyan tourism industry. Attending to one of the ethnic communities in Kenya, the Massai community, they reflected on the complex social positioning of women in this community that places them outside of the formal 'tourism industry' in ways that inhibit their ability to enhance the capacities of their handicraft businesses. With a lack of social policy support there are limited ways for these women entrepreneurs to (re)negotiate alternative entrepreneurial pathways, and thus, they remain positioned within an entrepreneurial context that is ultimately unsustainable.

Taken together these chapters highlight the disparate variations regarding the extent to which governments apply a social lens to entrepreneurship. In Peru and Colombia, for example, it appears that there have been important governance structures put in place to not only encourage but facilitate shifts towards higher numbers of women entrepreneurs in the tourism industry. Whilst in Kenya, where we already see high numbers of women in this space, a notable lack of support for such tourism entrepreneurs is evident. What is important to note in all three cases though is that there remains a lack of evidence on the *outcomes* of such social policy support, even when it is in place. A lack of policy evaluation is something that was similarly identified in Figueroa-Domecq, de Jong and Williams (2020) systematic literature review four years ago, and an issue that was only partly addressed in the recent Special Issue on gender, entrepreneurship and social policy within the *Journal of Sustainable Tourism*. This research gap has become of greater importance in more recent years, given the enhanced understandings regarding the vulnerabilities experienced by entrepreneurs in the tourism industry as a result of the Covid-19 pandemic. We thus here call for researchers working in this space to turn attention to outcomes of social policy interventions and the extent to which such outcomes are gendered.

An Interdisciplinary Turn for Tourism Entrepreneurship Research

Strong interest in qualitative approaches abound in the gender and entrepreneurship literature, facilitating rich insights into the individualised, lived experiences of entrepreneurs. At the same time, however, utilisation of social theory and philosophy to assist in making sense of such experiences remains limited and disjointed within the tourism and entrepreneurship literature (Figueroa-Domecq, de Jong and Williams, 2020). This is largely a response to the evolution of ideas within the tourism literature, more broadly, whereby research agendas have tended to be informed by business and managerial interests, leaving engagement with critical and socially informed ideas on the margins (Tribe, Dann and Jamal, 2015). Responding to the inconsistencies in uniting qualitative inquiry with social theory and philosophy, several chapters in this edited collection have engaged specifically with the feminist informed concept of performativity. Performativity as a concept was developed through the work of feminist philosopher Judith Butler (1990), whereby identities are understood as unstable, rather than innate, and come into being, and are constantly made, through everyday performance within specific spatial contexts. Engagement with the concept of performativity has enabled insights into the complex social issues affecting women entrepreneurs in the tourism industry, and the unique ways through which women entrepreneurs utilise their agency to negotiate such issues.

Informed by these ideas, Gizem drew on the concept of 'legitimacy-seeking' in Chapter 3 to help make sense of Turkish women's entrepreneurship experiences. Through a performative approach that recognises the fluidity and processual nature of entrepreneurial performance, the chapter explored the ways women entrepreneurs navigate the challenges of a patriarchal entrepreneurial landscape by either conforming to gendered entrepreneurial norms (doing gender) or challenging them (re-doing gender). The chapter further identified the discrepancies between the challenges women are required to negotiate as Turkish entrepreneurs and the lack of specific policy interventions designed to support such challenges, noting, for instance, that Turkey does not have any funded nursery school opportunities for women entrepreneurs nor any training opportunities for women seeking to undertake alternative forms of entrepreneurship, beyond the more traditional entrepreneurial practices of cooking, cleaning, and handicraft production.

Similarly drawing on a feminist-informed understanding of entrepreneurial performance in Chapter 7, de Jong, Ngoasong and Kimbu took focus on the accessibility of formal finance for entrepreneurs in Ghana and Nigeria. The chapter identified that whilst the governments of both countries are keen to support entrepreneurial tourism growth, particularly in the early stages of enterprise development, a two-tiered system prevails whereby women tended to be guided towards micro or small loans because of the gendered perceptions associated with women entrepreneurs and their enterprises, whilst men entrepreneurs had access to a broader range of formal finance that enabled them to acquire loans that better reflected their individualised circumstances. This two-tiered structurally embedded system prevailed despite entrepreneurs from all genders experiencing immense challenges in negotiating formal finance due to a lack of government and policy support in place to counter high interest rates, taxes and infrastructural costs. Both chapters point to the immense value in attending to individualised entrepreneurial performance when considering the social dimensions of entrepreneurship. A performative lens turns attention to the discrepancies between lived experience and government support and the ways such discrepancies can reproduce, if not exacerbate, existing gendered inequities.

The concept of performativity further informed Tankibayeva, MacDonald and Saduakassova's ideas in Chapter 5. This chapter examines the policy support for women entrepreneurs in Kazakhstan, detailing how Kazakhstani policy support is designed as multicomponent and cross-modular in terms of goals and time, facilitating women entrepreneurs to engage in ways that align with their individualised local context and life circumstances. The study further noted the crucial role frontline government staff undertake in communicating available support and shaping awareness and attitudes – particularly for those women located in rural areas. Such a multicomponent and cross-modular approach provides a potentially valuable model that could be applicable in other geographical contexts, given it is designed to prioritise individualised experience and positionalities. Importantly, however, for the participants in this study feelings of inexperience and anxiousness served to inhibit women's engagement with governmental support – again highlighting the importance of remaining attuned to the embodied and emotional dimensions of entrepreneurial experience, through concepts like performativity.

We are not the first to recognise that tourism scholarship has largely been constructed within the confines of business and management disciplinary thinking (cf. Nunkoo et al., 2021). Business and managerial approaches often sit within, and thus tend to adhere to, neoliberal ideals, and as such, these approaches are not always capable of rethinking the neoliberally informed constructions of the tourism industry. The confinement of tourism scholarship has thus led to the construction of distinctive disciplinary realms that work in parallel, inhibiting the development of ideas. We argue that it will be imperative in future research to engage more meaningfully with *inter*disciplinary approaches – particularly those within social theory and philosophy that are driven with a critical social imperative to bring to question the structural forces that inhibit entrepreneurial agency.

Rethinking Key Concepts in Response of Global Challenges

Various social theorical and philosophical ideas were further drawn on in several chapters to deconstruct and reconceptualise key concepts found within the tourism entrepreneurship literature. In Chapter 8, for example, Ngoasong took focus on the performance of women entrepreneurs during the Covid-19 pandemic, rendering insights into the ways women entrepreneurs negotiated the challenges of the pandemic. It was identified that rather than simply 'bouncing back' to the status quo, as has been the prevailing discourse and prioritisation of much policy response to the pandemic, some of the women entrepreneurs in this study also 'bounced forward', whereby they were able to embrace business transformation through innovation, improvisation and diversification.

This notion of 'bouncing forward' was utilised by Ngoasong to reconceptualise 'resilience'. The concept of resilience has tended to be drawn on in the business and managerial literature to consider the extent an individual or enterprise might possess the ability to 'bounce back', following a crisis. The Covid-19 pandemic, however, brought forth conditions wherein 'normal' ways of working and in consequence the ability to 'bounce back' was not available for extended temporalities, compelling the women entrepreneurs in this research to adopt new ways of working that transformed their enterprises in positive and lasting ways. Such redirections ultimately enhanced their understandings of themselves as resilient, in ways not previously imagined.

In Chapter 4, Carles, Fraiz and Vila took these ideas further by identifying the need for new concepts through which to make sense of shifting working conditions. More specifically, the chapter took focus on women housekeepers working in the Spanish hospitality industry, considering the unique performances adopted to confront the increasingly precariousness of their working conditions. The concept 'house-keepreneurship' was introduced as a lens to reflect the ways these individuals have drawn on entrepreneurial performances to construct a collective voice through the formation of the Las Kelly's association for women working as hotel housekeepers. This collective voice has assisted these individuals in (re)negotiating the precariousness of the working conditions in the Spanish services sector in ways that were not available to them as standalone workers.

Spain is not an anomaly in terms of the precarity of working conditions in the tourism industry; this is an issue felt internationally, right across the tourism industry. Given the widespread nature of these precarious working conditions, there is much to be learnt from the construction of the Las Kelly's association and the unique ways these individuals have been able to facilitate a collective voice that can respond to the shortcomings of Spain's economic, political and social systems. More broadly, global challenges are set to increase in their intensity and frequency. Chapters 4 and 8 thus importantly shed light on the ways how concepts, such as resilience and entrepreneurship, will themselves evolve as individuals and enterprises are forced to renegotiate such global challenges and the consequential effects on our everyday, working lives. It will become increasingly imperative for future research to remain attuned to such shifts and asses the ways these concepts take on new meanings.

Where to Now?

The chapters that constitute this edited collection are diverse, drawing on rich empirical insights, across a wide range of geographical contexts. This edited collection, however, provides a mere starting point for evaluating the potential of social policy within the context of gender and entrepreneurship. The chapters in this collection illustrate that whilst there are exceptions, the economic agenda prevails in consideration of the relation between policy and entrepreneurship – ensuring that women, and other marginalised communities, continue to experience greater challenges in their entrepreneurial journeys, a situation that persists despite the vulnerabilities made clear through the Covid-19 pandemic. As such, in this final chapter, we have hoped to identify three ways: future tourism entrepreneurship research might continue these discussions, to render a stronger research agenda that recognises the inequities experienced by entrepreneurs, particularly in regard to gender, and the role social policy undertakes in responding to such inequities. The ultimate impetus being to generate a body of critically informed, interdisciplinary research that can address, and communicate, the social and political needs of tourism entrepreneurs to academic, industry and policy networks.

References

Ahl, H., & Nelson, T. (2015). How policy positions women entrepreneurs: A comparative analysis of state discourse in Sweden and the United States. *Journal of Business Venturing, 30*(2), 273–291. https://doi.org/10.1016/j.jbusvent.2014.08.002

Butler, J. (1990). *Gender Trouble: Feminism and the Subversion of Identity*. New York: Routledge.

Crook, S. (2020). Parenting during the Covid-19 pandemic of 2020: academia, labour and care work. *Women's History Review, 29*(7), 1226–1238. https://doi.org/10.1080/09612025.2020.1807690

Figueroa-Domecq, C., de Jong, A., Kimbu, A., & Williams, A.M. (2024). Financing tourism entrepreneurship: a gender perspective on the reproduction of inequalities. *Journal of Sustainable Tourism, 32*(3), 656–676. https://doi.org/10.1080/09669582.2022.2130338

Figueroa-Domecq, C., de Jong, A., & Williams, A.M. (2020). Gender, tourism, and entrepreneurship: a critical review. *Annals of Tourism Research, 84*, 102980. https://doi.org/10.1016/j.annals.2020.102980

Henderson, K.A. (1994). Perspectives on analyzing gender, women, and leisure. *Journal of Leisure Research, 26*(2), 119–137.

Hmieleski, K., & Sheppard, L. (2019). The Yin and Yang of entrepreneurship: gender differences in the importance of communial and agentic characteristics for entrepreneurs' subjective well-being and performance. *Journal of Business Venturing, 34*(4), 709–730.

Kimbu, A., Ngoasong, M., & de Jong, A. (2024). Gender, entrepreneurship and social policy in tourism: tying the knot. *Journal of Sustainable Tourism, 3*, 421–437. https://doi.org/10.1080/09669582.2024.2309216

Nunkoo, R., Sharma, A., Rana, N., Dwivedi, Y., & Sunnassee, V. (2021). Advancing sustainable development goals through interdisciplinarity in sustainable tourism research. *Journal of Sustainable Tourism*. https://doi.org/10.1080/09669582.2021.2004416

Rogerson, C.M., & Baum, T. (2020). COVID-19 and African tourism research agendas. *Development Southern Africa, 37*(5), 727–741. https://doi.org/10.1080/0376835X.2020.1818551

Swart, M.P., Cai, W., Yang, E.C.L., & Kimbu, A.N. (Eds.). (2024). *Routledge Handbook on Gender in Tourism: Views on Teaching, Research and Praxis* (1st ed.). London: Routledge. https://doi.org/10.4324/9781003286721

Tribe, J., Dann, G., & Jamal, T. (2015). Paradigms in tourism research: a trialogue. *Tourism Recreation Research, 40*(1), 28–47. https://doi.org/10.1080/02508281.2015.1008856

UNWTO. (2020). *AlUla Framework for Inclusive Community Development through Tourism*. United Nations World Tourism Organisation.

Yang, E.C.L., & Ong, F. (2020). Redefining Asian tourism. *Tourism Management Perspectives, 34*, 100667. https://doi.org/10.1016/j.tmp.2020.100667

Index

Note: **Bold** page numbers refer to tables and *italic* page numbers refer to figures.

For Product Safety Concerns and Information please contact our EU
representative GPSR@taylorandfrancis.com
Taylor & Francis Verlag GmbH, Kaufingerstraße 24, 80331 München, Germany

www.ingramcontent.com/pod-product-compliance
Ingram Content Group UK Ltd.
Pitfield, Milton Keynes, MK11 3LW, UK
UKHW020002200926
475702UK00015B/2427

* 9 7 8 1 0 3 2 3 2 6 1 4 6 *